FLOOD FORECASTING USING ARTIFICIAL NEURAL NETWORKS

VRIJE UNIVERSITEIT
UNESCO-IHE Institute for Water Education

Flood forecasting using artificial neural networks

ACADEMISCH PROEFSCHRIFT

ter verkrijging van de graad van doctor aan
de Vrije Universiteit Amsterdam en
het UNESCO-IHE Intitute for Water Education te Delft,
op gezag van de rector magnificus
prof.dr. T. Sminia,
en de vice-rector prof.dr.ir. H. Savenije,
in het openbaar te verdedigen
ten overstaan van de gezamenlijke promotiecommissie
van de faculteit der Aard- en Levenswetenschappen
en het UNESCO-IHE Intitute for Water Education
op 20 augustus 2003 om 15.00 uur
in het auditorium van UNESCO-IHE

door

Pichaid Varoonchotikul

CRC Press
Taylor & Francis Group
Boca Raton London New York

CRC Press is an imprint of the
Taylor & Francis Group, an **informa** business
A TAYLOR & FRANCIS BOOK

promotor: prof.dr. M.J. Hall
copromotor: dr. A.W. Minns

First published 2003 by Taylor & Francis

Published 2007 by CRC Press
Taylor & Francis Group
6000 Broken Sound Parkway NW, Suite 300
Boca Raton, FL 33487-2742

ISBN-13: 978-90-5809-631-9 (pbk)

Visit the Taylor & Francis Web site at
http://www.taylorandfrancis.com

and the CRC Press Web site at
http://www.crcpress.com

To my parents, Kaejuer and Wanphen Varoonchotikul who try their best to let their children have the best education possible.

To Nopphakhun Somsin, who does not only encourage and support me, but also fought very hard for me to come and pursue my study here.

To the Thai people, who suffer from the economic crisis and still have been able to support my stay here.

Without those, I would never have been able to do this research.

Contents

Abstract

Flood disasters continue to occur in many countries around the world and cause tremendous casualties and properties damage. To mitigate the effects of floods, both structural and non-structural measures can be employed, such as dykes, channelisation, flood proofing of properties, land-use regulation and flood warning schemes. Sufficient advance warning time may save lives and property by allowing time to effect various structural and other adjustment. Earlier advanced warning can be achieved through advances in mathematical modelling. There are many rainfall-runoff models being developed and employed for flood forecasting which lead to the issue of flood warnings. Artificial intelligence techniques have recently been introduced, include the Artificial Neural Network (ANN) and the Support Vector Machine (SVM). Among these models, the Artificial Neural Network (ANN) has been found suitable for modelling the rainfall-runoff process in a wide variety of catchments under specific circumstances. An ANN is a quick and flexible approach which gives very promising results. Unfortunately, the inability to predict beyond the limits of the training range (extrapolation problem) was found to be a serious limitation of this approach. It is the solution to this problem that forms the one of the principal objectives of the present work.

The overview of forecasting methods is given to illustrate the wide variety of methods currently applied to the problems of flood forecasting. Then a general theory, concepts and the different network architectures of Artificial Neural Networks are deal with, especially their activation function values and derivatives. Preliminary considerations, such as a choice of a suitable time interval for analyses and a choice of goodness-of-fit indices, are given. A method of creating room for an ANN to extrapolate has been proposed, which focuses on the setting of the range of the activation functions applied to each node in the network. This approach has been successfully applied to several case studies, i.e. The Silk Stream and Rivers Thrushel and Brue. Different scale factors are also investigated and the results are summarised.

Recurrent Neural Networks, of both the Elman and the Jordan forms, are studied and compared to the basic Feed Forward Network. The results reveal that the Elman recurrent network offers no particular advantages over the Feed Forward Network, and both the Jordan recurrent network and the Feed Forward Network give comparable results. However, the Jordan recurrent network offers a possibility to recur the errors back to the context unit, which helps to improve the results. It is these recurred errors which leads to the further investigation of the *forward shift operator*. In an alternative approach, the First Derivative Transfer Function (FDTF) was employed to derive the forward shift operator that is input to the network to produce the forecast results. The use of ANNs to correct a crude forecast (the forward shift operator), demonstrated by the two examples of the Rivers Brue and Mole, has been shown to be more successful. The results reveal that the forward shift operator provides useful information which inhibit the phase errors as forecasting proceeds in time. In the case study of the River Mole, which lacked this information, the phase errors obtained with the tradition feed forward network had been pronounced.

Acknowledgement

I hereby express my great appreciation to

- The government of Thailand for providing financial support for my participation in the PhD programme at UNESCO-IHE.
- The Royal Irrigation Department of Thailand for approving paid leave for the entire period of my study in the Netherlands.

I wish to extend my sincere gratitude to

- Professor Dr. M. J. Hall for contributing his immense knowledge and experience in the field of research throughout this work and his patience during the author's difficulty periods.
- Dr. A.W. Minns for his constructive advice and comments' during this study.
- Dr. P.J.M. de Laat for supporting my attendance at the conference in China, and his encouragement, and friendly discussions during this study.
- Dr. Prof. Dr. J. P. O'Kane for his discussions and for drawing my attention to the FDTF technique which led to fruitful results.
- Dr. R.G.W. Venneker for his programming skills, friendly conversations and the translation of the Chapter 7 into the Dutch version in Chapter 8.
- Mr. Cheevin Limsiri and family for their personal and moral support, during my absences from home.

I am indebted to the staff of IHE, my colleagues and friends for their tremendous support for my studies as well as this long stay abroad apart from my family.

1

Introduction

1.1 Flood forecasting

According to the Worldwatch Institute, 1998 was a record year for economic losses arising from weather-related disasters, with a total global cost in excess of US$ 90 x 10^9. There is increasing evidence that losses are rising largely because more people are settling in areas that are prone to natural hazards, such as the flood plains and deltaic areas of major rivers. In the USA, for the period 1992-1996, floods accounted for 26 per cent of costs arising from natural disasters, amounting to some US$ 25 x 10^9 (van der Vink et al, 1998). The occurrence of a major flood remains a newsworthy item, whatever its geographical location. The incidence of several within one year, as occurred across Asia in 1995, is likely to invoke a degree of journalistic hyperbole, as exemplified by the headline *The Waters of Death* which appeared in the magazine *Newsweek* for 2 October, 1995.

Mitigating the effects of floods can be accomplished using either *structural* or *non-structural* measures, or by a combination of both. Structural measures consist of engineering works, such as dykes, channelisation or flood detention reservoirs, whose purpose is to change the shape (i.e. delay and attenuate the peak flow) of the flood hydrograph. In contrast, non-structural measures, such as the flood proofing of property, land-use regulation within flood plains and flood warning schemes, are intended to reduce the economic losses in a flood situation. The last two decades have seen a growing appreciation that the traditional structural approaches can engender complacency within the population that they protect, such that even larger economic losses occur on the rare occasions on which the design standard of the works is (inevitably) exceeded. Indeed, the *World Bank Policy Paper on Water Resources Management* (World Bank, 1993: 60) proposed that:

"Governments should intensify efforts to achieve flood control with non-structural measures that are less costly, yet no less effective in preventing (these) disasters, than more expensive structural measures".

These sentiments echo those of Nemec (1986: 2), who when discussing *Hydrological Forecasting Systems (HFSs)*, i.e. systems for the prior estimation of hydrological phenomena in real time, noted that:

A flood-oriented HFS serving well-established, disaster prevention operations should, in most cases, prove more efficient in mitigating the effects of major floods than would structural measures.

In discussing HFSs, a distinction should be made between the sequential processes of (1) producing a *forecast* of the hydrological variable, i.e. flood stage or discharge; and (2) comparing the forecast with predetermined threshold values, which if exceeded result in the issue of a *warning* to the general public. The ability to forecast hydrological variables depends upon the inherent time lag between the cause (precipitation) and the effect (streamflow), which is capable of representation by some form of computer model. A typical classification of models for the purposes of hydrological forecasting, adapted from that provided by Nemec (1986; pp 74-75) can be summarised as follows:

A. Purely deterministic forecasting procedures (models)

1. Hydrometric data-based (involving only streamflow data)
   a. correlations between stages or discharges at downstream and upstream sites on the same river system
   b. hydrological (storage) routing
   c. routing with a computational hydraulic model

2. Rainfall/runoff models
   a. regression models of streamflows on rainfall totals and indices of catchment wetness
   b. systems (transfer function) models, with either *lumped* (averaged) or *distributed* precipitation inputs
   c. conceptual soil moisture accounting models

B. Hybrid stochastic-deterministic forecasting procedures

1. prescriptive time-series models
2. deterministic models with stochastic updating procedures

Models in each of the above categories vary according to their data requirements and the computational resources that are necessary for their implementation. The methodology adopted may also vary with the problem to be addressed. For example, one of the simplest approaches to flood forecasting is the direct plotting of the peak stages (or discharges) at a downstream gauge against those of an upstream gauge in the same catchment (A.1.a). Examples of such a relationship for the Han River, Korea and the Sacramento River in California, USA were given by Lee (1967, Fig. 3) and Linsley et al (1982; Fig. 9-17), respectively. An allowance may be made for the time of travel between the two points, and an equation may be fitted to the plotted data using simple linear regression analysis. This method obviously will be confined to relatively large catchments, and for points between which there is a proportionally small increase in contributing area. For smaller catchments, a related technique, which was favoured in the then US Weather Bureau in the 1940s and 1950s, was the coaxial diagram. As illustrated by Linsley et al (1949; Chapters 16 and 22), such diagrams were a form of nomographs in which the user entered information on the independent variables, and then moved parallel to the major axes, variable by variable, to estimate the value of the independent variable. In a typical example, Bruce and Clark (1966; Figure 11.1) present a coaxial diagram in which total storm runoff for Southern Ontario is estimated from the rainfall total, an antecedent precipitation index (API) and a week number. Such relationships can, of course, be reconstructed simply as multiple linear regression equations.

For large river basins having long travel time, routing procedures based upon measured upstream flows (A.1.b and A.1.c) are probably the preferred technique. They are more flexible than a simply stage or discharge correlation (Franchini & Lamberti, 1994). Advances in computational hydraulics from the 1970s onwards have resulted in the development of commercial software for the solution of river routing problems. Such models are capable of modelling both converging and diverging river networks in great detail, but can be very demanding in terms of data on river cross sections, which should in many cases extend far beyond the river channel, and channel roughness among other parameters. The complete dynamic models are based upon the numerical solution of the De Saint Venant equations (see Abbott and Minns, 1998, for a thorough treatment of this topic). However, approximation techniques, such as the diffusion analogy, have been applied successfully for flood routing problems (see, for example, Price, 1973, for applications of approximation techniques to British rivers). A further approximation, known as the Kinematic Wave, has also been applied extensively in flood routing, but in the better known form of the Muskingum method. In effect, the latter approach allows the direct computation of the outflow hydrograph from the inflow hydrograph for an arbitrary length of reach. In its original form (e.g. Chow, 1959; pp 609-616), the Muskingum method makes no allowance for lateral inflow along the reach, but a three-parameter version introduced by O'Donnell (1985) has been successfully applied in New Zealand (O'Donnell et al, 1988) and China (Yang, 2001). These techniques are particularly appropriate to catchments that are large enough for travel times from an upstream to a downstream gauging station to provide sufficient time to issue a forecast. Otherwise, forecasts must be based upon the direct estimation of flows from observed (or forecast) precipitation. The hydrological routing procedures (A.1.b) do not give consistently satisfactory results, perhaps due to being unable to cope with a river reach that has undergone erosion and changed its physical characteristics, or overbank flow. This problems lead to other approaches, for example the rainfall-runoff models.

There are many general references on rainfall-runoff modelling, for example Singh and Frevet (2002a,b). The rainfall-runoff models vary from lumped conceptual models e.g. Tank model (Sugawara, 1995), to distributed conceptual models e.g. ARNO model (Todini, 1996), and the physically based distributed model e.g. SHETRAN (Mellor et al, 2000).

When the catchment response time is short, which is the case in small and medium sized catchments, rainfall-runoff models are probably the preferred techniques. The regression model of streamflow on rainfall total and indices of catchment wetness (Sittner, 1969) is the improvement over the models in category A.1.a, but it increases model complexity. The Unit Hydrograph (UH) is another approach (A.2.b) widely used in the field of Hydrology (Sherman, 1932; Nash, 1957; Dooge, 1959, 1973; Sempere Torres et al, 1992; Duband, 1993). Kothyari & Singh (1999) employed a distributed UH-concept (A.2.b) to the Narmada catchment, India, by sub-divided it into smaller sub-areas based upon the hydro-physiographical homogeneity indicated by vegetation, slope, soil type and the long-term rainfall isohyetal maps. In applying the UH, one needs to know the effective rainfall which is the result after the total rainfall has undergone complex processes inside the catchment. To simplify those complex processes one can, for example, introduce the constant lost rate, the API-index and the transmission loss function (Sharma & Murthy, 1995). A further approach is to apply the concept of Soil Moisture Accounting (A.2.c), which replicated these processes (Ford, 2001; Cameron, 2002). In the case of inhomogeneous catchments, the distributed rainfall-runoff models are preferred (Garrote & Bras, 1995; Todini, 1996;Takeuchi, 2001)

Section (B) of the classification by Nemec (1986) introduces a distinction between *prescriptive models*, which provide statistical resemblance relative to a vector of parameters, and *descriptive models*, in which the parameters retain a physical interpretation. Time-series models based upon the family of Autoregressive Integrated Moving Average Processes, as introduced by Box and Jenkins (1976), are perhaps the most widely used examples of prescriptive models. For flood forecasting purposes, such models emphasise the strength of serial correlation between flows at successive time intervals that is inherent in flood hydrographs. A related approach is that of reconstructing the hydrological modelling problem in a so-called *state-space formulation*, such as the following:

$$X_{t+1} = A.X_t + B.U_t + e_t \quad (1.1)$$

$$Y_{t+1} = C.X_t + v_t \quad (1.2)$$

where the U_t and Y_t are vectors of system inputs and outputs respectively; X_t is a vector of system states representing causality and internal structure; *A, B* and *C* are weighting matrices; and e_t and v_t are vectors of noise, generally assumed to be Gaussian with zero mean and finite variance. Given the 'noisy' model of the state dynamics of the hydrological system and the definition of a 'noisy' output or measuring system represented by Equation (1.2), a method for developing an 'optimal' forecast, i.e. a forecast with minimum variance, is required. The solution adopted by many investigators has been the *Kalman Filter algorithm* (see Wood and O'Connell, 1985; Bras and Rodriguez-Iturbe, 1985, Chapter 8). Examples of such a hybrid approach for the hypothetical river was given by Shiiba (2000), and for forecasting floods at the Kagami dam catchment, Japan by Nagai (2002).

However, short warning lead times limit the damage reduction that can be attributed to forecasting (Ford, 2001). To increase the lead time forecast, Lardet & Obled (1994) applied a stochastic rainfall generator in order to extend the lead-time of flood forecasts on small catchment named the Gardon d'Anduze and finally concluded that the rainfall forecasting models are required to improve the flood forecast. Rainfall forecasting can be achieved by using radar (Bressand, 2002; Deshons, 2002) and a high-resolution climate model (Christensen & Christensen, 2003). Kim & Barros (2001) used a Quantitative Precipitation Forecasting (QPF) model for improved flood forecasting in four catchments on the leeward side of the Appalachian mountains in the mid-Atlantic region of the USA. The model employed both Numerical Weather Prediction (NWP) output and rainfall and radiosonde data. Jasper et al (2002) applied a coupling atmospheric and hydrological model to the Lago Maggia basin of 2,627 km^2 in Switzerland in order to improve flood forecasting in the Alpine area. Pedder et al

(2000) carried out the short period forecasting of rainfall in the River Brue catchment and pointed out the limitations of an operational NWP, and its sensitivity to initial conditions. Bell & Moore (2000) employed a simple two-dimensional rainfall forecasting model to forecast rainfall input to the IH PDM (Probability Distributed Moisture) model, to assess its effect on flow forecast accuracy in the River Brue. New approaches are possible through radar, coupling a QPF with a rainfall-runoff model. However, such systems require large investment and unlikely to be implemented in developing countries for some years.

Although there are many models used in flood forecasting, some are not able to cope with dynamic changes inside the catchments. Some models are sensitive to the initial state of the catchments (Garrote & Bras, 1995), some models are too complicated to calibrate and need to have a robust optimisation tools (Duan et al, 1994), and some models require the understanding of the physical processes inside the basin (Garrote & Bras, 1995). These problems lead to exploration of a more data-driven approach.

The recent introduction of Artificial Intelligent tools, for example, the *Artificial Neural Networks (ANNs)* and the Support Vector Machine (SVM), to the field of water resource has opened various new approaches in modelling. Both are data-driven models. The ANNs are recognised as being universal approximators and are capable of extracting the underlying relationship between any input, or stimulus, and its subsequent output, or response. The ANNs compose of layers with many nodes in each layer. Input data are fed into the first layer called the input layer, while the outputs are taken from the last layer, called the output layer, and the layers in between are hidden layers. To employ the ANNs, the network architecture must be chosen beforehand (see Chapter 2 for more details). Many examples of applying ANN-models were given by Hall & Minns (1993); Minns & Hall (1996); Campolo et al (1999a); Dawson & Wilby (1998) and Cameron (2002). The recent research on ANNs as hydrological models has offered the opportunity to avoid prior processing, for example the *effective rainfall* (total rainfall minus losses), since ANNs are not constrained by volume continuity. Nor does a complex physical structure have to be established prior to modelling. The ANNs offer not only a relatively quick and flexible means of modelling (Imrie et al, 2000), but also a way to find relationships between different input samples (Dawson & Wilby, 1998). In contrast, the SVM is basically a linear machine, which can be seen as a statistical tool that approaches the problem in a similar way to an ANN. The architecture of a SVM does not have to be determined before training and the modelling function may be chosen among a great variety of possible options. Moreover, SVMs are trained by solving a constrained quadratic optimisation problem, which implied that there is a unique optimal solution for each choice of the SVM parameters. This is essentially different from the ANNs trained using back propagation. Examples of applying SVM models to simulate flows in a catchment in UK was given by Han & Yang (2001), and to simulate flow in Tryggevælde catchment in Denmark by Yu et al (2002).

Although papers describing interest in the application of ANNs as hydrological models only began to appear in the early 1990s, a considerable body of literature has now accumulated. Since comprehensive reviews have been provided by ASCE Task Committee (2000a,b), Maier and Dandy (2000), Dawson and Wilby (2001) and Minns and Hall (2003), only a brief overview of these techniques will be given here.

There are three broad categories of problems that are involved in the use of ANNs as general modelling tools, and their particular employment for forecasting purposes:

I. The overall architecture of the neural network, and the algorithms required to obtain the values of their weights;
II. The ability of the neural network to extrapolate beyond the range of its training data set; and
III. The selection of the inputs to the ANN that are required to represent the 'pattern' which gives rise to the 'pattern' of outputs.

Although the advantages of different architectures of network in specific applications are acknowledged, the intention here is to concentrate on the use of Multi-layer Perceptrons (MLPs) whose weights are obtained with the back-propagation algorithm. However, attention will also be paid to the efficacy of both *Jordan* and *Elman* networks in which the predicted outputs are fed back into the input layer either from the output layer (Jordan) or the hidden layer (Elman).

The solutions to the second and third of these problems are fundamentally dependent upon the application of the modeller's *domain knowledge*, and form the principal focus of this study. Minns and Hall (2003) have noted that the first choice to be made by the ANN modeller is the mode of presentation of the data to the network, i.e. how can the input and output patterns to be defined? One possibility is to take the *n* successive ordinates of the rainfall hyetograph and feed these into the *n* input nodes of a network whose *m* output nodes carry the *m* successive ordinates of the flow hydrograph (Lange, 1998). An alternative strategy is the so-called *dynamic approach* in which the input is a set of concurrent ordinates of (say) the rainfall totals from *p* raingauges within the catchment and the output is the concurrent rate of outflow. In this mode, each time step defines a pattern, and therefore a single storm event provides as many exemplars as there are runoff ordinates, rather than only a single input-output pairing. This is the approach that has been adopted by the majority of writers on rainfall-runoff modelling using ANNs, but requires a much higher level of hydrological insight into the working of the catchment system. Within the dynamic approach, three types of model can be identified:

The naïve dynamic model;

$$Q(t) = f(r(t), r(t-1), \ldots) \tag{1.3}$$

The rainfall-runoff simulation model;

$$Q(t) = f(r(t), r(t-1), \ldots, Q(t-1), Q(t-2), \ldots) \tag{1.4}$$

The 'auto regressive' model;

$$Q(t) = f(Q(t-1), Q(t-2), \ldots) \tag{1.5}$$

where $Q(t)$ is the outflow at time level t and $r(t)$ is the rainfall ordinate at time level t.

The simplest, naïve dynamic rainfall-runoff model (1.3) would consist of an ANN with inputs from one or more raingauges at time t, and an output of concurrent flow. A simple scatter plot of input(s) against output is sufficient in this case to indicate that the description of the input pattern is inadequate. Some improvement is obtainable by allowing for the time lag between the occurrence of the flow and the incidence of the causative rainfall. Since the flow at any instant is effectively composed of contributions from different sub-areas whose time of travel to the outlet covers a range of values, both the concurrent and antecedent rainfalls can be considered to be contributing to the outflow. Use of a moving *window* of rainfall at time *t* and the *k* previous intervals provides some improvement. The choice of length for the moving window of rainfall can significantly affect the accuracy of the resulting ANN model. If the window is too short, the input data does not contain enough information about the entire rain event that is contributing to the concurrent outflow. Conversely, if the window is too long, the input contains too much information, which may even include historical rainfall events whose effects have long since passed out of the catchment and so are no longer contributing to the concurrent flow.

A further problem with model (1.3) is that the simplistic input patterns may easily result in ambiguous results. More specifically, intervals with zero rainfall inputs are encountered in two contrasting situations: immediately prior to the beginning of the storm and the start of the rising limb; and some time after the end of the storm event when flows are moderately high and in recession. The ANN has no information to discriminate between these two 'no-rainfall' conditions and once more becomes 'confused'. These conditions have to be differentiated by the addition of another input if the ANN is to achieve the correct mapping. The most obvious candidate is a flow ordinate, which is most easily provided by the output at time *t* to become an input at time *t+1* (Hall & Minns, 1993; Minns & Hall, 1996; and similar use of stage outputs by Campolo et al, 1999a). This approach is described above as model (1.4). In effect, the flow (or stage) ordinate is employed as a crude measure of catchment wetness, although other researchers prefer to use a moving average of flow and/ or rainfall as an alternative measure of wetness (Dawson & Wilby, 1998).

Model (1.5) is then the logical extension of models (1.3) and (1.4) into purely 'auto-regressive' time series prediction. This model does not make use of any rainfall data at all but uses only antecedent outflow values as input to the ANN to predict the concurrent outflow. This ANN model provides only slightly poorer performance than the results from model (1.4), but suffers from an inability to capture the timing of peak flows. At the peak, the network has no other information available that tells it at which level the rising limb should stop until the actual measurements indicate that this is so. A similar *phase error* occurs at the beginning of the recession limb, caused by the ANN having no knowledge about the cessation of the rainfall until one or two time steps after the actual measured values start to decrease.

In general, the literature involving the application of ANNs to rainfall-runoff modelling confirms the exceptional accuracy of ANN models for short forecasting intervals. Longer forecasting intervals may be obtained by utilising *Q(t+1), Q(t+2),...* etc. as outputs during the training of the ANN. Unfortunately, the performance of the ANN decreases quite rapidly with an increasing prediction time horizon (Campolo et al., 1999a,b; Zealand et al., 1999). Another approach is to use a trained ANN with a 'feedback' loop in which the predicted output is used directly as input data for the subsequent time step. Unfortunately, the error accumulation associated with this approach also means that the performance of the ANN deteriorates quite rapidly after only one or two iterations (van den Boogaard et al., 1998). The more promising approach would appear to involve the use of *(partial) recurrent neural networks*, which contain feedback loops in both the training and recall modes of the network using the Jordan/Elman architecture. Proaño et al. (1998) and van den Boogaard et al. (1998) show significant improvements in long-term predictions using recurrent neural networks, which they rather unfortunately refer to as 'auto regressive' neural networks.

1.2 Aims of this study

The focus of this study is essentially the three problems already described in Section 1.1 above, but interpreted in the context of the application of ANNs to flood forecasting. Despite the nature of neural networks as *universal approximators* (Hornik et al, 1989) and *pattern recognition devices* (Aizerman et al, 1964a,b) there remains a fundamental need to apply ANNs in the context of *domain knowledge* about the problem being addressed. Item (III) above is therefore of fundamental importance to the whole project. A satisfactory solution to the extrapolation problem of item (II) is a prerequisite to successful forecasting, and will therefore be addressed first. The third item concerning network architecture and weight determination has received attention in many fields of application, and apart from exploration of the possibilities offered by recurrent networks, this problem receives lesser attention in this study. The following notes provide an indication of the overall directions of the project.

1.2.1. The extrapolation problem

The fact that a neural network is *"a prisoner of its training data set"* (Minns and Hall, 1996) has all too often been ignored by those who have applied ANNs to environmental problems. The result is usually a network that cannot exceed the extremes contained in its training data set when faced with higher or lower extremes in the testing data set. A variety of approaches have previously been suggested to overcome this problem, including

- modelling first differences in output rather than the original ordinates
- inclusion of a notional *Estimated Maximum Flood* (EMF) in the training data set
- manipulation of the range of standardisation of inputs and outputs to prevent saturation of the activation functions applied at each node of the network

The successful use of an EMF in training has been reported by Hettiarachchi et al (2003; under review), and limited success with first differences has been reported by Minns and Hall (1997).

The new approach that is suggested here to manage this extrapolation problem is to deal directly with the standardising range of input data. It involves setting the minimum and the maximum values of the possible range of input data to known values right at the beginning. This approach has been found to work well on the basis of selected examples and case studies. More details are presented in Chapter 4.

1.2.2. Network architecture

Once the extrapolation problem has been managed, the suggested approach will be applied throughout the rest of the work. The advantage of using recurrent networks over the feed forward networks is explored in Chapter 5. The potential for application of Jordan/Elman ANN architectures for flood forecasting requires further investigation. Initial studies have demonstrated that a sustained high level of performance can be achieved with a *moving time frame* type of forecasting system, that is broadly equivalent to a MLP with exogenous inputs type of recurrent structure. This and other simple recurrent network structures are investigated.

1.2.3. Choice of inputs

Two types of flood forecasting application are to be addressed: the flood routing problem, i.e. the hydrometric data-based category of Nemec (1986); and the rainfall-runoff modelling problem. Representative data sets for each application have been obtained from both Europe and the Far East (see Appendix I). Since the purpose of the project is to assist in the development of operational procedures, the selection of input data has to be confined to precipitation and streamflow time series, and to derived variables that are developed from such data, i.e. accumulations over moving time frames, spatial and temporal averages, etc.

Particular attention must be paid to the size of the data interval relative to the notional response time of the catchment under study. For example, daily time series are of little value in forecasting the behaviour of a catchment whose time-to-peak is 24 h or smaller. The size of the catchment is also an important consideration, since response time reduces with drainage area with a consequent reduction in the available lead time for a flow forecast. In an extreme situation, a precipitation forecast may be necessary instead of measurements of rainfall. Although forecasts of large scale precipitation up to ten days ahead are made globally on a routine basis, their pixel size of one degree of latitude by one degree of longitude limits their utility to large catchments (Hu, 2002).

Following a review of the available literature, the simple formula for the most appropriate sampling time for rainfalls and discharges suggested by Sugawara (1995) was found to provide acceptable results for these studies (see Chapter 3). If too large a sampling interval is assumed, then the smoothing of the variations in both the precipitation inputs and the flow outputs can lead to a deterioration in forecasting performance. An equally important issue in rainfall-runoff modelling is the choice between lumped (areally-averaged) and distributed precipitation inputs. If the catchment is divided into sub-areas, each with its own input streams of concurrent and antecedent rainfall totals, then the size of the ANN in terms of numbers of weights is greatly increased. The benefits of employing distributed rainfall inputs are therefore investigated more fully in Chapter 6.

As mentioned previously, phase error is a prevailing problem with ANN rainfall-runoff models. When measured variables, such as precipitation and runoff, are used as input variables up to a time t, then a phase error with a minimum value of one time unit can occur in the forecast of flow at time t+1. What the ANN requires is some indicator as to whether the output will reach its peak value within the next time interval. Such an indicator can be provided by employing an alternative procedure to make an initial forecast for time t+1, and then to use the ANN to correct (and improve) that forecast. This approach has been investigated by employing the First Difference Transfer

Function (FDTF) method of Duband et al (1993) as the initial forecast procedure. Further details are presented in Chapter 6.

1.3 Thesis outline

This work is composed of six chapters. A short overview of the material to be presented in the following chapters is given here.

Chapter 2 describes the biological neuron and the artificial neuron or perceptron. The feed-forward network is introduced along with its learning rule and various activation functions. The recurrent network algorithms are then described.

Chapter 3 describes the preliminary considerations that are necessary for the study. They concern the choice of a suitable time interval and the goodness-of-fit criteria to be applied in the modelling. The effects of choosing different sampling time intervals on the rainfall-runoff modelling process are demonstrated.

Chapter 4 describes the extrapolation problems inherent in applying artificial neural networks as rainfall runoff models. The way to manage these problems is then elaborated, illustrated by a variety of examples.

Chapter 5 describes the results of application of the recurrent neural networks to the modelling of the rainfall-runoff process and to the flood routing process.

Chapter 6 is concerned with the choice of input to a neural network applied to the modelling of the rainfall-runoff process. A comparison is made between the use of distributed and areal average rainfall as network inputs. The FDTF method is then introduced as a means of providing a forward shift operator for managing the phase errors. Results are illustrated by two contrasting examples.

Chapter 7 summarises the results obtained and the conclusions drawn from the previous chapters.

The Appendix I provides a general description of the data used in this study

2

Artificial Neural Networks

2.1 Biological Neuron

From neurophysiology, it is known that the human brain contains a massively interconnected net of 10^{10} to 10^{11} neurons or nerve cells. Each of these neurons is connected to 10^3 to 10^4 dendrites further down. A very simplified schematic view of a biological neuron is shown in Fig. 2-1. A typical cell has three major regions: the cell body, which is also called the *soma*, the *axon*, and the *dendrites*.

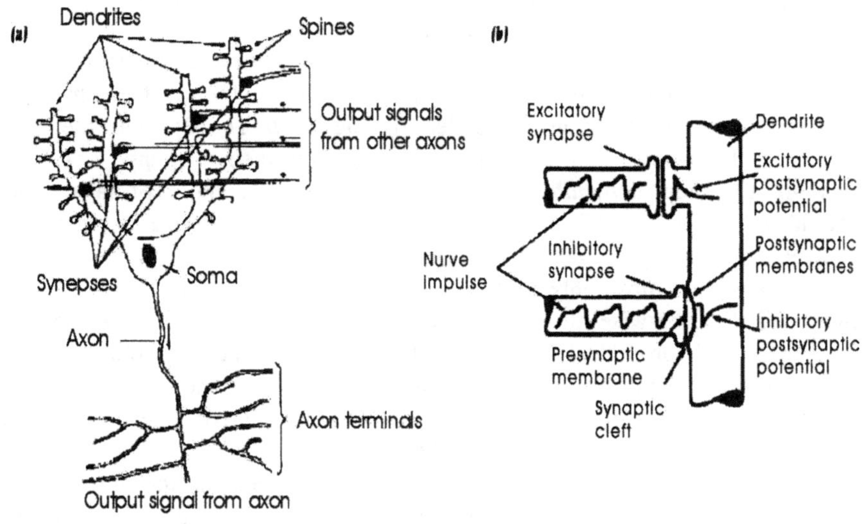

Figure 2-1 (a) Schematic structure of a biological neuron (b) Simplified sketch of synapses (Adapted from Cichocki & Unbehauen, 1992).

The *soma* or *body cell* is the large, round central body in which almost all the logical functions of the neuron are realised. It receives signals in the form of electrical pulses that have originated from other neurons from a tree-like network of nerve fibres call *dendrites*. There are 10^3 to 10^4 dendrites per neuron which connect the neuron to a set of other neurons. After the soma processes the incoming signal or information, it may produce an output signal depending on the strength of the incoming signal compared to a certain threshold level. Once the output signal is produced, it is distributed into various branches, which make *synapses* on to the dendrites and soma of other neurons, via the *axon*. The axon is usually highly branched and is a nerve fibre attached to the soma. Synapses are the termination points for the axons and play the role of interfaces connecting some axons of the neurons to the spines of the input dendrites.

2.2 What is an Artificial Neural Network

Artificial Neural Networks (ANNs) have been motivated right from their inception by the recognition that the human brain computes in an entirely different way from the conventional digital computer. The brain is a highly complex, non-linear, and parallel computer (information-processing system). It has the capability to organise its structural constituents, know as neurons in a massively distributed and parallel network, so as to perform certain computations (e.g., pattern recognition, perception, and motor control) many times quicker than the fastest digital computer in existence today.

ANNs consist of a large number of simple processing elements called neurons or nodes. Each node is then connected to other nodes by means of direct links. Each link is associated with a weight that represents the strength of outgoing signal. The processing of each node is carried out in two steps, that is, the weighted sum of the inputs is taken, and is followed by the application of the activation function.

ANNs are usually implemented by using electronic components or are simulated in software on a digital computer. They are characterised by (1) their patterns of connections between the neurons (called its *architecture),* (2) their methods of determining the weights on the connections (called their *training,* or *learning, algorithm),* (3) their *activation function,* and (4) their number of layers: single (Hopfield nets); bilayer (Carpenter/Grossberg adaptive resonance networks); and multilayer (most back-propagation networks).

If all the signals flow only in one direction, an ANN is called a Feed-Forward Network (FFN). Otherwise, they are called Recurrent Networks. The most commonly used ANNs in hydrological rainfall-runoff models are Feed-Forward Multilayer Perceptrons (FFMLPs), which usually contain three layers and employ monotonic activation functions in all layers. Their weights are usually obtained by supervised training using the back-propagation algorithm.

The learning ability and performance of an ANN depends on the suitability of its architecture which is needed to be pre-specified, more specifically, the number and configuration of its hidden nodes (Imrie et al., 2000). A type of a FFN that avoids the need to manually locate the optimum network architecture is called the cascade-correlation learning architecture (Fahlman & Lebiere, 1991), which constructively builds the network by adding one hidden node at a time. There are a number of successful applications of this algorithm in hydrological modelling (Karunanithi et al., 1994; Muttiah et al., 1997; Durucan and Imrie, 1998).

Another alternative is the Radial Basis Function network (RBF), which is also a FFMLP but employs a non-monotonic activation function. For example, the Gaussian activation function is used in the hidden layer and a linear activation function in the output layer (Norgaard et al., 2000). It also has a different learning algorithm that is carried out in two phases, first for the hidden layer and then for the output layer (Mason et al., 1996). The hidden layer is self-organising; its parameters depend on the distribution of the inputs, not on the mapping from the input to the output. The output layer, on the other hand, uses supervised learning in order to set its parameters. Mason et al. (1996) and Dibike (1997) applied RBF networks to the problem of rainfall-runoff modelling. Dibike (1997) demonstrated that the training time of RBF networks was usually significantly less than for equivalent FFMLP networks when dealing with only small numbers of input data sets.

2.3 Multilayer Perceptron

The rainfall-runoff model used in this study is based on a Feed-Forward Multilayer Perceptron (FFMLP). A typical three-layer feed forward network is shown in Figure 2-2.

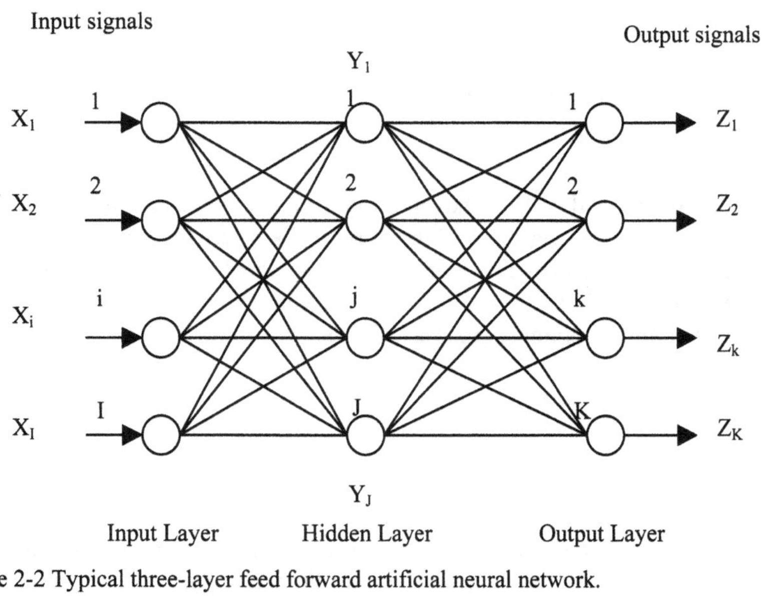

Figure 2-2 Typical three-layer feed forward artificial neural network.

The first layer, called the *input layer*, consists of I nodes and connects with the input variables. This layer performs no computation but is used to distribute I inputs into the network. The last layer connects to the output variables and is called the *output layer*. This layer consists of K output nodes. The one or more layers of processing units located between the input and output layers have no direct connections to the outside world (the output variables) and are called *hidden layers*. An MLP may have a number of hidden layers. However, only one hidden layer will be considered here. This layer consists of J hidden nodes.

The processing elements in each layer are called *neurons, units, cells,* or *nodes.* Each of the nodes is connected to all the nodes of neighbouring layers. The parameters associated with each of these connections are called weights. These weights represent information being used by the net to solve a problem. In general, all connections are 'feed forward'; that is, they allow information transfer only from an earlier layer to the next consecutive layers. Nodes within a layer are not interconnected, and nodes in non-adjacent layers are not connected, which is not the case for *recurrent neural networks.* The recurrent neural network will be considered further in Section 2.6.

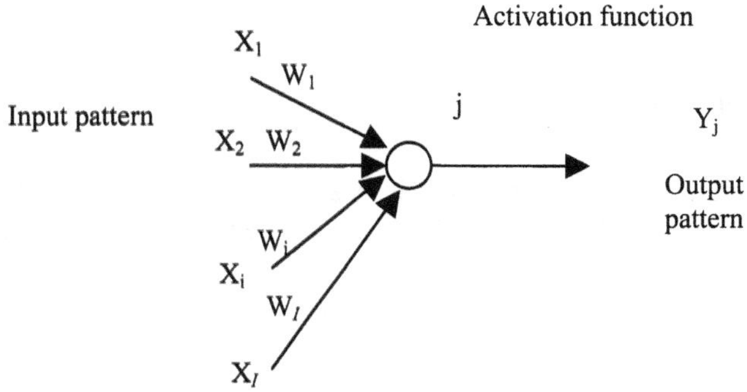

Figure 2-3 A typical ANN node with I input signals.

The architecture of a typical node is shown in Figure 2-3. Each hidden node j receives I incoming signals (x_i), from every node i in the previous layer (for example from Input layer to Hidden layer). Associated with each incoming signal (x_i), there is a weight (w_{ij}) connected between layer I and J. The effective incoming signal (net_j) to node j is the weighted sum of all the incoming signals as follows:

$$net_j = \sum_{i=1}^{I} w_{ij} x_i \qquad (2.1)$$

This effective incoming signal (net_j) passes through a *nonlinear activation function* (sometimes called a *transfer function* or *threshold function*) to produce the outgoing signal, called the *activation* or *activity level,* (y_j) of the node. Typically, a neuron sends its activation as a signal to several other neurons. It is important to note that a neuron can send only one signal at a time, although that signal is broadcast to several other neurons simultaneously.

2.4 Delta learning rule for Feed-Forward Multilayer Perceptron

The delta learning rule is also called the *LMS (Least Mean Square), error back propagation training* algorithm. Input patterns are submitted sequentially during the feed forward propagation training. If a pattern is submitted and its classification or association is determined to be erroneous, the synaptic weights as well as the thresholds are adjusted so that the current least mean square classification error is reduced (Schalkoff, 1997). Therefore, the purpose of this algorithm is to adjust the synaptic weights w_{ij} (i=1,I) and w_{jk} (j=1,J), which connect Input layer to Hidden layer and Hidden layer to Output layer, respectively, to assure a minimisation of the error function (E_k). The instantaneous estimate of the mean-square error calculated at the output layer is defined as follows (Rumelhart et al, 1986):

$$E_k = \frac{1}{2}(z_k - t_k)^2 \tag{2.2}$$

where z_k is the output from output layer and t_k is the target value. The outgoing signal (z_k) is a function of the activation as follows:

$$z_k = f(net_k) \tag{2.3}$$

The effective incoming signal net_k, comprising weighted sum signals from the hidden layer (y_j) is calculated as follows:

$$net_k = \sum_{j=1}^{J} w_{jk} y_j \tag{2.4}$$

The outgoing signal (y_j) is the result of the incoming signal net_j, being passed through the activation function, which is not necessarily the same as that for z_k, as follows:

$$y_j = f(net_j) \tag{2.5}$$

Again, the effective incoming signal net_j, comprising weighted sum signals from input layer (x_i) is calculated as:

$$net_j = \sum_{i=1}^{I} w_{ij} x_i \tag{2.6}$$

The derivative of the error function with respective to w_{jk}, can be expressed after applying the chain rule, as follows:

$$\frac{\partial E_k}{\partial w_{jk}} = \frac{\partial E_k}{\partial z_k} \frac{\partial z_k}{\partial net_k} \frac{\partial net_k}{\partial w_{jk}} \tag{2.7}$$

The relevant terms are substituted into (2.7), resulting in the formula:

$$\frac{\partial E_k}{\partial w_{jk}} = (z_k - t_k) f'(net_k) y_j \tag{2.8}$$

The derivative of the error function with respective to w_{ij} is as follows:

$$\frac{\partial E_k}{\partial w_{ij}} = \left[\frac{\partial E_k}{\partial z_k} \frac{\partial z_k}{\partial net_k} \frac{\partial net_k}{\partial y_j} \right] \frac{\partial y_j}{\partial net_j} \frac{\partial net_j}{\partial w_{ij}} \tag{2.9}$$

Substituting all relevant terms into (2.9) results in the formula:

$$\frac{\partial E_k}{\partial w_{ij}} = [(z_k - t_k) f'(net_k) w_{jk}] f'(net_j) x_i \tag{2.10}$$

The weight adjustment of both w_{ij} and w_{jk} are based on the gradient descent search, which changes the weights in the direction in which the error surface goes down most steeply, as follows:

From input layer to hidden layer:

$$\Delta w_{ij} = -\eta \frac{\partial E_k}{\partial w_{ij}} \tag{2.11}$$

Where η is a learning parameter.

Substituting (2.10) into (2.11) then

$$\Delta w_{ij} = -\eta [(z_k - t_k) f'(net_k) w_{jk}] f'(net_j) x_i \tag{2.12}$$

From hidden layer to output layer:

$$\Delta w_{jk} = -\eta \frac{\partial E_k}{\partial w_{jk}} \tag{2.13}$$

Substitute (2.8) into (2.13) then results in

$$\Delta w_{jk} = -\eta (z_k - t_k) f'(net_k) y_j \tag{2.14}$$

Finally, the updated weights are

$$w_{ij} = w_{ij} + \Delta w_{ij} \tag{2.15}$$

$$w_{jk} = w_{jk} + \Delta w_{jk} \tag{2.16}$$

Both (2.12) and (2.14) are very much dependent on the derivative of the activation function employed in the network. This fact also suggests that training speed is sensitive to the activation function via its derivative.

2.5 Activation functions

There are various activation functions employed in ANNs, the most commonly used ones being the following the unipolar binary function or sigmoid function (S), the bipolar binary function (B), the hyperbolic tangent function (T) and the linear function (L):

2.5.1 The unipolar binary function or sigmoid function (S)

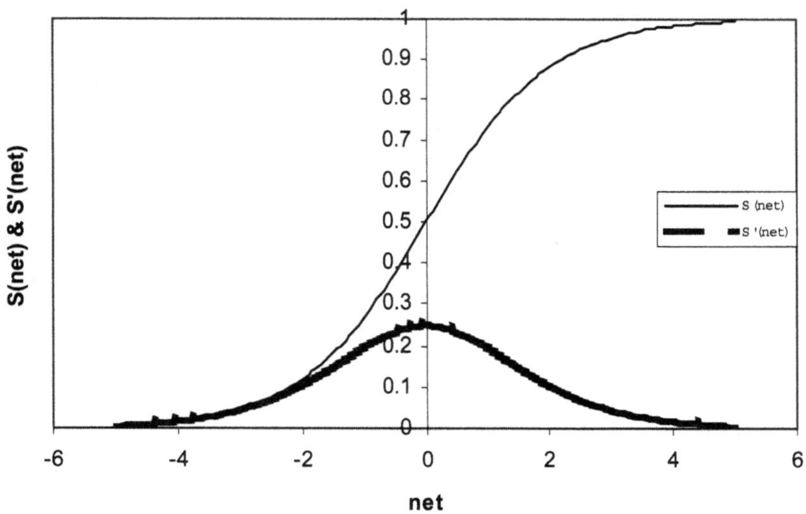

Figure 2-4 Sigmoid function and its derivative.

The sigmoid function, as implemented by Hall & Minns (1993), Dawson & Wilby (1998), and Campolo et al (1999), is defined as

$$S(net_j) = \frac{1}{1 + \exp\left(-\lambda net_j\right)} \tag{2.17}$$

where $\lambda > 0$ in (2.17). λ is proportional to the neuron gain determining the steepness of the function, which in general is set equal to one. The term λnet_j can vary on the range

$\pm\infty$, but $S(net_j)$ is bounded between (0 and 1). The derivative of the sigmoid function is as follows:

$$S'(net_j) = \lambda S(net_j)(1 - S(net_j)) \tag{2.18}$$

The sigmoid function and its derivative are shown in Figure 2-4. As the derivative approaches zero then λnet_j tends to $\pm\infty$. At the maximum value of 0.25λ, λnet_j is equal to zero and its function value is 0.50. The derivative of the sigmoid function is almost constant and varies between 0.22 to 0.25 when the function falls within the range of (0.3 to 0.7) which corresponds to the λnet_j values of (-0.847 and +0.847).

2.5.2 The bipolar binary function (B)

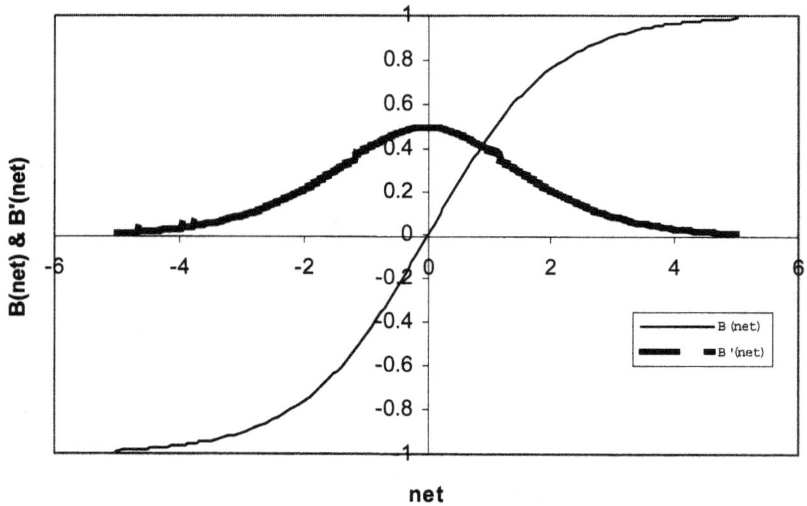

Figure 2-5 Bipolar binary function and its derivative.

The bipolar binary function can be obtained by shifting and scaling the sigmoid function as follows:

$$B(net_j) = 2S(net_j) - 1 \tag{2.19}$$

Substituting (2.17) into (2.19), then results in

$$B(net_j) = \frac{1 - \exp(-\lambda net_j)}{1 + \exp(-\lambda net_j)} \tag{2.20}$$

where $\lambda > 0$ in (2.20). Again, λ is proportional to the neuron gain determining the steepness of the function, which in general is set equal to one. The term λnet_j can vary on the range $\pm\infty$, but $B(net_j)$ is bounded between (-1 and 1). The derivative of the bipolar binary function is as follows:

$$B^{'}(net_j) = \frac{\lambda}{2}(1 - B(net_j))(1 + B(net_j)) \tag{2.21}$$

The bipolar binary function and its derivative are shown in Figure 2-5. Its derivative is at a maximum value of 0.50λ when λnet_j is equal to zero, and $B(net_j)$ is related to the hyperbolic function as shown below:

$$B(net_j) = \tanh\left(\frac{\lambda}{2} net_j\right) \tag{2.22}$$

The bipolar binary function is therefore a subset of the hyperbolic tangent function.

2.5.3 The hyperbolic tangent function (T)

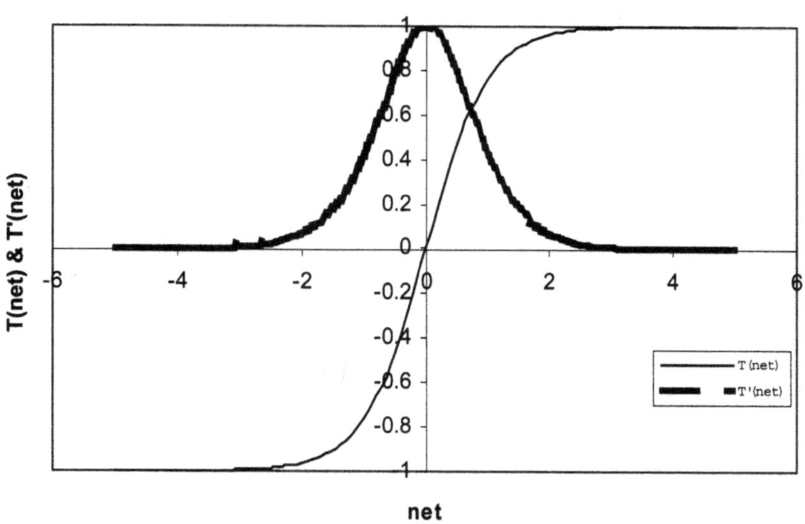

Figure 2-6 Hyperbolic tangent function and its derivative.

The hyperbolic tangent function, as implemented by Varoonchotikul et al (2002b), is defined as follows:

$$T(net_j) = \tanh(\lambda net_j) \tag{2.23}$$

where $\lambda > 0$ in (2.23). Once more, λ is proportional to the neuron gain determining the steepness of the function, which in general is set equal to one. The term λnet_j can vary on the range $\pm\infty$, but $T(net_j)$ is bounded between (-1 and 1).
The derivative of hyperbolic tangent function is as follows:

$$T'(net_j) = \lambda(1 - T(net_j))(1 + T(net_j)) \tag{2.24}$$

The hyperbolic tangent function and its derivative are shown in Figure 2-6.The value of the gain factor (λ) in the hyperbolic function is twice the value of the bipolar binary function. As a result, the derivative of hyperbolic function, which equals λ, is much steeper than that of the bipolar binary function and the sigmoid function.

2.5.4 The linear function (L)

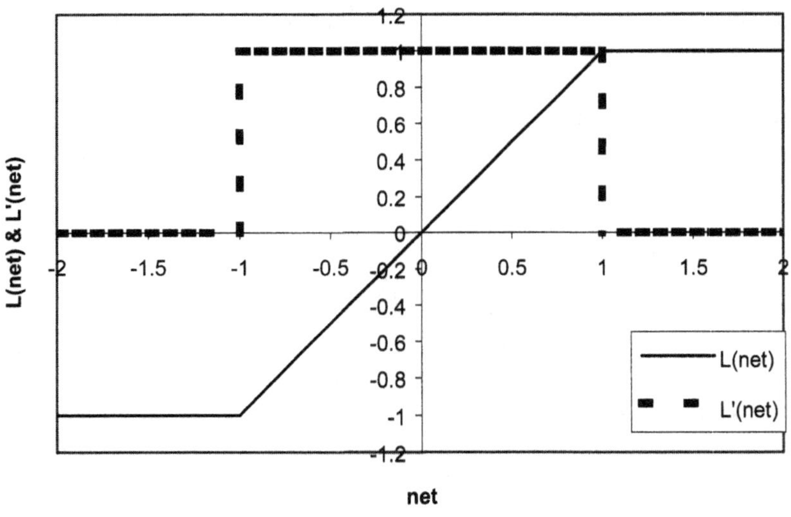

Figure 2-7 Linear function and its derivative.

The linear function, as implemented by Karunanithi et al (1994), is defined as follows:

$$L(net_j) = \lambda net_j \tag{2.25}$$

where $\lambda > 0$ in (2.25). λ is proportional to the neuron gain determining the steepness of the function, which in general is set equal to one. The term λnet_j can vary on the range $\pm\infty$, but $T(net_j)$ is bounded between (-1 and 1). The derivative of the linear function is:

$$L'(net_j) = \lambda \tag{2.26}$$

Both linear and hyperbolic functions have their maximum derivative values equal to λ at their centroid. They are much steeper than those of the bipolar binary function and the sigmoid function. The linear function and its derivative are shown in Figure 2-7.

2.6 Recurrent Multilayer Perceptron

Recurrent Multilayer Perceptron (RMLP) networks are most conveniently introduced by beginning with a simplified presentation of the FFMLP network. Figure 2-2 can be schematised into a block diagram as shown in Figure 2-8. Starting from the input layer, an array of input (X_i) is fed to the next layer, which is the hidden layer. The net effective incoming signals (X_i) produce net_j and the output signal Y_j to the next layer (the output layer). These net effective incoming signals (Y_j) produce net_k and finally, output signal Z_k. By comparing this output signal Z_k to the target T_k, the error back propagation then starts to adjust the network weights to minimise the error function.

The recurrent scheme differs from the FFMLP in the way that its output recurs either from the output layer or the hidden layer back to its input layer. The two different schemes are described in the following sections.

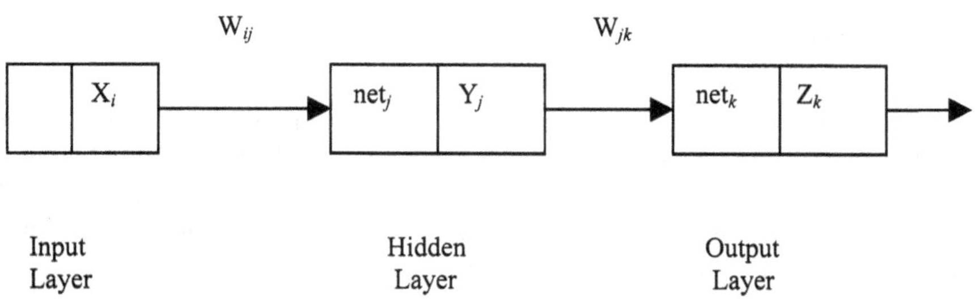

Figure 2-8 Schematic diagram of Feed Forward MLP.

2.6.1 Elman Recurrent Network (ERN)

Elman (1990) suggested the architecture shown in Figure 2-9, which is now known as the *Elman Recurrent Network*. The network employs its output signals from the *hidden layer* to train the network. The input layer is divided into two parts: the true input units and the context units, which are a tapped-delay-line memory of q units from the hidden layer.

The modifiable connections are all feed forward, and the weights from the context units can be trained in exactly the same manner as the weights from the input units to the hidden units by conventional back propagation methods (Fausett, 1994). This is done at each time step in the case of input sequences. The context units are treated like inputs in applying back-propagation.

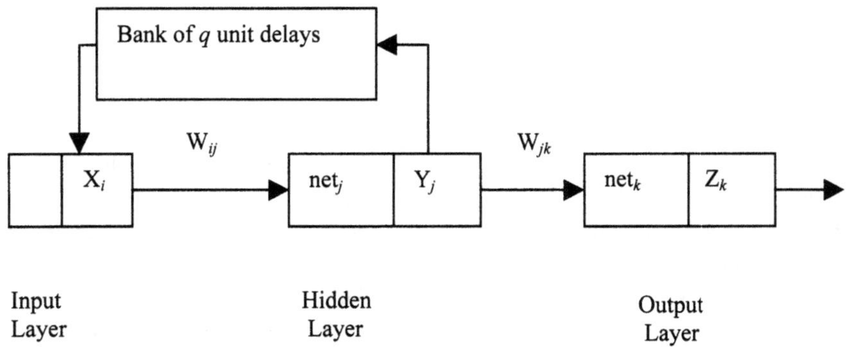

Figure 2-9 Elman Recurrent Network

2.6.2 Jordan Recurrent Network (JRN)

Jordan (1986, 1989) suggested a different recursive scheme now known as the *Jordan Recurrent Network.* The network employs its output signals from the *output layer* to train the network. Its output signals are fed back to the input layer via context units (tapped-delay-line memory of q units). Again, the weights from the context units can be trained in exactly the same manner as those of ordinary type *ERN* (Fausett, 1994).

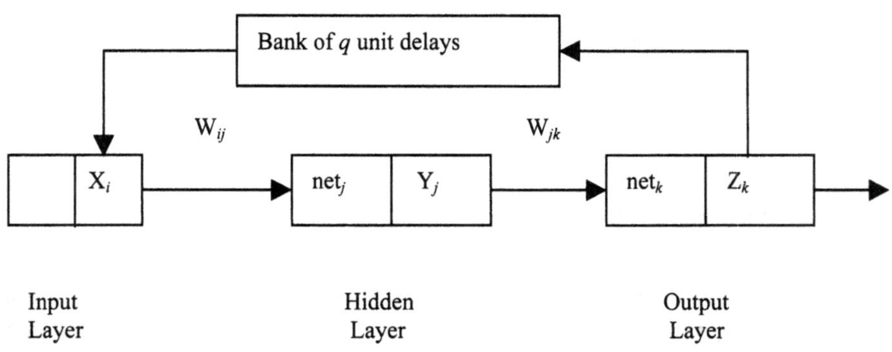

Figure 2-10 Jordan Recurrent Network

3

Preliminary considerations

3.1 Introduction

In any rainfall-runoff modelling exercise, there are two important practical considerations: the choice of a suitable time interval (T_s) at which to sample the storm events on which the modelling exercise is based; and the selection of appropriate goodness-of-fit indices on which to judge the efficacy of the derived model(s). The importance of choosing a suitable sampling interval is perhaps not always appreciated. The majority of hydrological data bases (for example, the FRIENDS* European Water Archive) contain daily data. However, for many catchment areas of (say) less than 1000 km^2, this time interval is too long to capture the essence of the response to storm rainfall. In spite of these limitations, attempts have been made to regionalise the parameters of daily rainfall-runoff models (Wheater et al, 2002, for example). The data interval problem is discussed more fully in section 3.2. With regard to goodness-of-fit of computed to observed results, modellers have tended to rely on a limited number of simple indices to describe their results, perhaps the most popular of such indices has proved to be the so-called Coefficient of Efficiency (CE) introduced by Nash and Sutcliffe (1970). Both the null hypothesis upon which this index is based and its general behaviour over its range of variation (plus one to minus infinity) have attracted some criticism (Beran, 1999; Hall, 2001). The problems of goodness-of-fit are outlined in greater detail in section 3.3.

3.2 The choice of a suitable time interval

An important consideration in any rainfall-runoff modelling exercise, which perhaps has not received the attention that it deserves, is the choice of the time interval at which the data are extracted. In general, a time interval, T_s, should be adopted that captures the appropriate level of variability in the time series of precipitation and the streamflow. The empirical rules proposed for the division of the rising limb of the finite-period unit hydrograph into six (Snyder, 1938; as explained by Johnstone and

* : Flow Regimes from International Experimental and Network Data Sets.

Cross, 1949) or five time steps (NERC, 1975) may be seen as attempts to accomplish this objective. By comparison, the use of a daily time unit for a catchment of 66.2 ha for the ANN modelling study performed by the ANNEXG Group (2002) was more of a non-linear regression exercise than a test of rainfall-runoff modelling. More formal approaches, based on the concept of the Nyquist frequency, by Childs (1982; see Chapman, 1994) resulted in the following expression for the time interval to be adopted:

$$T_s = \frac{\pi Q_{\max}}{6S} \tag{3.1}$$

where Q_{max} is the maximum flow ordinate and S is the maximum slope (positive or negative) of the flow hydrograph. According to Chapman (1994), this equation gives a time interval of approximately 5 minutes for a catchment of 40 ha, and 3 hours for a drainage area of around 600 km^2.

In a further contribution, Sugawara (1995) proposed that for small, steeply sloping drainage basins in Japan,

$$T_s = 0.05\sqrt{DA} \tag{3.2}$$

where T_s is in hours and *DA* is the drainage area in km^2. This equation tends to provide lower estimates than that of Childs, e.g. around 2 minutes for an area of 40 ha and 1.25 h for 600 km^2. In general there are other catchment properties which significantly influence the time step, for example, geomorphology, urbanisation, and slope. Therefore this formula may not be applicable everywhere in the world. Since previous work (Varoonchotikul et al, 2002a) had shown that Equation (3.2) provided a reasonable basis for one-step-ahead forecasting, this expression was adopted for further work.

Based on this criterion, the catchments under study are classified into three groups as very detailed (VD), moderately detailed (MD) and roughly detailed (RD). The first category has a measurement interval T_m less than or equal to the T_s, the second one is close to T_s (greater than T_s and less than or equal to twice the T_s) and the last one is much greater than T_s (greater than twice the T_s). The T_s-values of each river basins used in this study are tabulated in Table 3.1.

The Feed Forward Network (FFN), applied to data for which the sampling or measuring time interval (T_m) is less than or equal to T_s, performed best in one time step forecasting. For demonstration purposes, the Brue data were resampled into different intervals from 30 minutes to 1, 2, 3, 4, 6, 12 and 24 hours. These data were then input to a FFMLP to predict one time step ahead. Table 3.2 shows that, with higher ratios of (T_m/T_s), the forecast results deteriorate markedly. The coefficient of efficiency (CE; see Section 3.3) is higher than 99% when the ratio of (T_m/T_s) is less than two. Beyond that the CE dropped rapidly in a non-linear way. This behaviour is caused by the lagged data not providing enough information to the network, and there is less correlation among the data of the longer sampling times. Finally, the CE reaches a constant value, which is equivalent to the correlation of baseflow. Both the root mean square error (RMSE) and the mean absolute error (MAE) values are increased as the ratio of (T_m/T_s)

is increased. Therefore, the data used should have a sampling time less than or equal to twice T_s to guarantee the best results.

Except the Rivers Brue and Mole that their data are supplied on a 15 min interval, the rest of the catchments are on hourly basis. Therefore, no discussion is made on these catchments.

Table 3.1: Appropriate sampling time interval (T_s) of various catchments being studied in this thesis.

Catchment Name	DA (km^2)	Group	T_m (hr)	T_s (hr)	Ratio T_m/T_s
Brue River	135	VD	0.25	0.58	0.43
Chumporn River	1,819	VD	1.00	2.13	0.47
Dart River	248	MD	1.00	0.79	1.26
Dollis Brook	23.99	RD	1.00	0.24	4.17
Mole River	142	VD	0.25	0.60	0.42
Silk Stream	31.25	RD	1.00	0.28	3.57
Thrushel River	112.7	MD	1.00	0.53	1.89

Remark: DA : drainage area or catchment area; VD : very detailed ($T_m/T_s \le 1$); MD : moderately detailed ($1 < T_m/T_s \le 2$); RD : roughly detailed ($2 < T_m/T_s$);.

Table 3.2: Results of feed forward networks on River Brue data with different sampling time intervals to forecast one-step ahead.

T_m & Forecast Time (hr)	T_m/T_s	Epochs	CE	RMSE	MAE
0.50	0.86	8833	99.918	0.135	0.111
1.00	1.72	8259	99.428	0.228	0.093
2.00	3.45	5546	95.025	0.675	0.262
3.00	5.17	6986	88.100	1.045	0.317
4.00	6.90	5692	79.480	1.368	0.422
6.00	10.34	6026	55.406	2.076	0.669
12.00	20.69	5753	37.699	2.093	0.992
24.00	41.38	6151	37.551	2.251	1.135

Remarks: CE : the coefficient of efficiency; RMSE : Root mean square of error; MAE : Mean absolute error.

3.3 The choice of goodness-of-fit indices

At present, there are many rainfall-runoff models including lumped and distributed, physically based and conceptual, linear and non-linear, etc. that are employed in the field of Hydrology. In general, these models transform the input data e.g. rainfall, evaporation, upstream discharge, etc. into the output e.g. discharge. The question arises as to which features of the computed and observed outputs should be emphasised in determining the efficacy of the model. To answer this question, there are a number of indices that describe various goodness-of-fit of the models, but none of them is a universal index (Diskin & Simon, 1977). Some of the indices are listed in Table 3.3, which is adapted and expanded from that presented by Hall (2001). As an example, the maximum absolute error is concerned with the peak discharge while the total mean square error in monthly volume is for the volumetric error, etc.

Among these indices, the coefficient of efficiency (CE) is the most widely adopted in the field of Hydrology because of its ease of use in the process of calibration. Modellers tend to have a general appreciation of the relative performance of the model based upon such single measures (Hall, 2001). As a consequence, they tend to recognise it as the coefficient of determination (CD), which is wrong. Both CE and CD are correlation-base measures but the CE ranges from plus one to minus infinity while the CD ranges from plus one to zero, with higher values indicating better agreement. When both have the same value, e.g. 0.70, the interpretation is different. In the case of CD, this value means that the model can explain 70% of the variance in the observed data, while in the case of CE the squared differences between the observed and model simulated values is 30% of the variance of the observed data (Legates & McCabe, 1999).

Beran (1999) has criticised the null hypothesis implicit in the structure of the CE that tends to exaggerate the significance of the hydrograph fit and hence to exaggerate the real level of modelling skill. Beran also suggested that the modeller should not be satisfied with a CE lower than the mid-to-high 90s in percentage terms. Furthermore, Legates & McCabe (1999) have concluded that correlation and correlation-based measures should not be used to assess the goodness-of-fit of a hydrologic model and that additional evaluation measures e.g. RMSE, MAE should supplement model evaluation.

Hall (2001) has conducted numerical experiments on a sequence of model outputs subject to three different types of errors: volume error, bias and timing error. The results showed that negative values of the CE can occur both in the presence of bias in computed outputs, and when the computed volume of flow greatly exceeds the observed volume of flow. The errors in timing and magnitude of peak will lower the values of CE and either lead or lag timing of peak will give same CE. The author also mentions that the CE lacks discrimination for cases close to perfect reproduction and suggests the use of a coefficient based upon the first differences of the data to increase the sensitivity.

In this work, the errors in timing of the rising peak are referred to as *phase error* and are analysed by direct visualisation, no index has been calculated. Although the CE is not a perfect goodness-of-fit index, it will be utilised throughout this study. This index is widely accepted and assists in the comparison of the studies to the previous works of Hall & Minns (1993), Minns & Hall (1996, 1997) and Dawson & Wilby

(1998). However, use of CE is complemented by consideration of RMSE and MAE as appropriate.

Table 3.3 Goodness-of-fit measures for a typical daily rainfall-runoff model (Adapted from Hall,2001)

Feature	Definition
Root mean square error (RMSE)	$\sqrt{\frac{1}{\mathrm{n}}\sum_{i=1}^{n}(q_{Oi}-q_{Ci})^2}$
Total mean square error in monthly volume	$\sum_{i=1}^{n_{month}}\left(\frac{1}{\mathrm{n}_{\mathrm{day}}(i)}\sum_{k=1}^{n_{day}(i)}(q_{Oi}-q_{Ci})\right)^2$
Mean absolute error (MAE)	$\frac{1}{\mathrm{n}}\sum_{i=1}^{n}\lvert q_{Oi}-q_{Ci}\rvert$
Maximum absolute error	$\max\ \lvert q_{\mathrm{Oi}}-q_{Ci}\rvert;\ 1\leq \mathrm{i}\leq \mathrm{n}$
Coefficient of efficiency (CE)	$1-\frac{\frac{1}{\mathrm{n}}\sum_{i=1}^{n}(q_{Oi}-q_{Ci})^2}{\frac{1}{n}\sum_{i=1}^{n}(q_{Oi}-q_B)^2}$
Maximum difference in the highest peak discharge	$\max(q_{\mathrm{Oi}})-\max(q_{\mathrm{Ci}});1\leq \mathrm{i}\leq \mathrm{n}$
Mean error in time-to-peak for all storm events	$\frac{1}{n_e}\sum_{j=1}^{n_e}(T_{Oj}-T_{Ej})$
Root mean square error in time-to-peak	$\sqrt{\frac{1}{n_e}\sum_{\mathrm{j}=1}^{\mathrm{n_e}}(T_{Oj}-T_{Ej})^2}$
First serial correlation coefficient	$\frac{\sum_{i=1}^{n-1}(q_1-q_B)(q_{i+1}-q_B)}{\sum_{\mathrm{i}=1}^{\mathrm{n}}(q_i-q_B)^2}$
Coefficient of determination (CD)	$\frac{\sum_{i=1}^{n-1}(q_1-q_B)(q_{Ci}-q_{Bc})}{\left[\sum_{\mathrm{i}=1}^{\mathrm{n}}(q_i-q_B)^2\right]^{0.5}\left[\sum_{\mathrm{i}=1}^{\mathrm{n}}(q_{Ci}-q_{Bc})^2\right]^{0.5}}$

Notation : q_{Oi} and q_{Ci} represent the observed and computed flows for day i, $1 \leq i \leq n$; q_B and q_{Bc} represent the mean of the observed and computed flows; $n_{day}(i)$ is the number of days in a month; n_{month} is the number of months in the time series; n_e is the number of storm events in the observed and computed series; T_{Oj} and T_{Ej} are the observed and estimated times-to-peak of the jth storm event

4

Extrapolation management for Artificial Neural Network models of Rainfall-Runoff relationships

4.1 Introduction

Flood disasters have often occurred in many countries due to urbanisation, deforestation, climatological change, etc. They cause impacts to society that go beyond cost and facilities, including family and community disruptions, dislocation, injuries and unemployment. In order to cope with these problems efficiently for public safety and water management, flood forecasting is necessary. Hydrologists and engineers have utilised rainfall-runoff models for a long time, successfully to some extent. The models vary from simple rational formulae, simple black box models, conceptual models, etc. to physically-based models e.g. SHE (Abbott et al, 1986). The more complex the model, the larger the input data and the computational time required by the model. One drawback is that the costs of setting up and operating a complex model can be high and an expert is required to operate such a model.

The Artificial Neural Network (ANN) approach has been found capable of identifying complex, non-linear relationships between the rainfall and runoff in a wide range of catchments. This new approach has a number of advantages over the conventional hydrological models, since it deals with the observed values of inputs and outputs without applying unrealistic assumptions and simplifications on ever-changing complex natural hydrological systems. The ANN "learns" the complexity of the natural systems by adjusting its weights based only on the data supplied for the training and can make very accurate predictions. It is flexible and can be used as a lumped or a distributed model according to the requirements. It is simple in data requirements, economical, computationally fast and easy to apply.

Many researchers [for example, Hall & Minns (1993); Zhu et al (1994); Minns & Hall (1996, 1997); Shamseldin (1997); Campolo et al (1999 a, b); Dawson and Wilby (1998)] have shown that an Artificial Neural Network is capable of simulating runoff from rainfall. Once the ANN is properly trained, the computational time is small. One difficulty is that the ANN is a "prisoner of its training data" because the predicted values will not be greater than the maximum trained value put into the ANN (Flood & Kartam, 1994; Minns & Hall, 1996). It is generally accepted that ANN cannot extrapolate beyond the range of the training data (Maier & Dandy, 2000; Bowden et al, 2002). Minns & Hall (1996) also point out that there is a chance that future rainfall will be larger than the historic records. Therefore the ANN will not be able to extrapolate this flood because its magnitude will be larger than those of the events contained in trained data.

This identifies the need to overcome this extrapolation problem if the ANN is to be operated in real-time flood forecasting. Imrie et al (2000) suggested a way to solve the problem by adding a "guidance system" to the cascade correlation architecture instead of the more commonly used multi-layer perceptron. In fact, this approach indirectly uses the desirable flow value as a model calibration and may not be applicable to different catchments or even more extreme hydrological events. Hettiarachchi et al (2003) had another approach in which the estimated maximum flood (EMF) in the river basin was computed and added to the training data for the ANN. This approach seems to work quite well but it still depends on how well the EMF value and its hydrograph is estimated.

Kaastra & Boyd (1995) and Karunanithi et al (1994) thought that it could be an advantage to employ the sigmoid activation function in the hidden layer and the linear activation function in the output layer when it is necessary to extrapolate beyond the range of the training data. Carrieré et al (1996) employed *bipolar linear normalizations* in the input layer, and the *logistic function* was used for activation in the nodes of the hidden and output layer. However, the *sigmoid activation function* appears to have been the most widely used. Therefore, the extrapolation problems will firstly be managed according to this function (see also Varoonchotikul et al, 2002a) and then extended to the rest.

Although the data sampling time intervals will affect the results of the ANNs as mentioned in Chapter 3, no experiments at different time intervals were carried out in this chapter because all data used were supplied on an hourly basis.

4.2 Standardisation : key to managing the extrapolation problem

Since the *sigmoid* function (equation 2.17) used in the back-propagation algorithm varies between 0 and 1, it is prudent to standardise all input values before passing them into a neural network. Without this standardisation, large values input into an ANN would require extremely small weighting factors to be applied. The most common standardised range is between (0.1 to 0.9) which is not necessarily a good choice. The *sigmoid* function varies its values in a non-linear way, conforming with the S-shape, which means that those values located near the end of the range will approach upper and lower limit saturation and be transformed back to their original values in a distorted way. Therefore, those raw data should preferably be standardised into the range of (0.3 to 0.7) since the function is almost linear within this interval as previously pointed out in section 2.5.1. The back transformation will restore all outputs in a proportional manner, which is necessary.

Owing to the output range of the *sigmoid* function, all values leaving an ANN are automatically output in a standardised format. These output values must be "*destandardised*" to provide meaningful results. This can be achieved by simply reversing the standardisation algorithm used on the training data in the output nodes.

In order to manage the extrapolation problem, the possible minimum and maximum value of the raw data (*MINV* and *MAXV*) are estimated. The range of data is then calculated according to

$$RANGE = (K2 * MAXV - MINV) \tag{4.1}$$

The factor *K2* is introduced to provide the room for the possible maximum value to be accommodated in this range, and its choice will be clarified later in this Chapter. At this point, the *K2* is set to *2* pending the discussion in section 4.8. Since there is also a chance that the low value will be lower than the recorded one, it is preferred to provide some clearance of *1%* of the *RANGE* in the transformation.

Let *MAXT* and *MINT* be the upper and lower limits of the maximum and minimum value of raw data to be transformed. Then they are calculated as follows

$$MAXT = (K2 * MAXV + \frac{RANGE}{100}) \tag{4.2}$$

$$MINT = (MINV - \frac{RANGE}{100}) \tag{4.3}$$

Finally, the raw data D_i is transformed to neuron input X_i, which will vary in the range of (0.3 to 0.7) as follows

$$X_i = 0.3 + 0.4 * \left[\frac{D_i - MINT}{MAXT - MINT} \right] \tag{4.4}$$

4.3 Standardised range of other activation functions

Other activation functions described in section 2.5 are the *bipolar binary function* (B), the *hyperbolic tangent function (T)* and the *linear function (L)*. Since the *bipolar binary function* relates to the *sigmoid function* according to equation (2.19), its standardised range should also follow the same relation. This brings the standardised range to (-0.4 to 0.4). As mentioned before, the bipolar binary function is a subset of the *hyperbolic tangent*, therefore the same standardised range shall be applied to the *hyperbolic tangent function*.

The standardisation steps suggested in equations (4.1)-(4.3) are the same, but the transformation must be changed to the ranges suggested above. Therefore, in case of the *biploar binary function* and *hyperbolic tangent function*, equation (4.4) is changed to:

$$X_i = -0.4 + 0.8 * \left[\frac{D_i - MINT}{MAXT - MINT} \right] \tag{4.5}$$

In case of the *linear function*, no specific measures are required, but if one plans to mix it with other activation functions in the network, the standardised range should be the same as the others so as not to cause confusion when back transformation takes place.

4.4 Case studies without including the EMF in the raw data

To test the applicability of the proposed methodology, it was applied to previously studied data such as those of Minns & Hall (1997), Minns (1998), and Hettiarachchi et al (2003). Four catchments in the UK were the focus of these studies. These catchments were the Silk stream, the Dollis brook, the Thrushel and the Dart. Descriptions of these catchments are presented in Appendix I.

In the previous studies, the authors employed the *Sigmoid activation function* for both hidden and output layers, and the calculation was performed until the global error of the network was a minimum. Those authors standardised these catchment data sets between 0.1 and 0.9, while 0.3 and 0.7 are employed in this study. Minns & Hall (1997) and Minns (1998) noted that the global error, as used in most software packages, was also a function of the number of nodes in the network, which might not be a good basis for comparison. Therefore they preferred to present their results in terms of the well-known coefficient of efficiency *CE* (Nash and Sutcliffe, 1970) defined as follows:

$$CE = 1 - \frac{\sum_{i=1}^{N} \left(Q_i - Q_{cal,i}\right)^2}{\sum_{i=1}^{N} \left(Q_i - \overline{Q}\right)^2} \qquad (4.6)$$

where the Q_i are the N measured flows with mean $\overline{Q}$ and the $Q_{cal,i}$ are the N calculated flows.

In this study, the *sigmoid activation function* was employed for both hidden and output layers, and the results are also present in terms of *CE* for purpose of comparison.

4.4.1 Silk Stream catchment

For the Silk Stream data, Minns & Hall (1997) and Minns (1998) took 51 storms in chronological order and selected the first 13 and the last 10 storms to be included in the training data set, with the remaining 28 reserved for verification. This mixture of storms not only conveniently encapsulated the historical changes in percentage impervious area, but also ensured that the maximum flow in the training data of 9.8 m^3/s was smaller than the peak of 13.8 m^3/s in the verification data. They constructed the network with eight inputs (rainfall at time *t*, and the four previous time steps, and the flows at three previous time steps) and one output of flow at time *t*. There were three layers in the network with four nodes in the hidden layer. The network structure was therefore 8-4-1.

The results revealed that the extrapolation problem was apparent in the case of the Silk Stream, as shown in Figure 4-1. As expected, the maximum calculated flows were not larger than 9.8 m^3/s, which was the maximum flow in the training data.

By utilizing the same network configuration (8-4-1), the new concept of standardisation was implemented. The values of *MINV*, *MAXV*, *MINT* and *MAXT* for the rainfall and flow are summarised in Table 4.1 along with values for the other three catchments.

All data were transformed to the range of (0.3 to 0.7) according to equation (4.4) and input to the network. The results obtained from the network were destandardised and are shown in Figure 4-2. The coefficient of efficiency was 0.969. There is no problem with extrapolation in this case.

Table 4.1: Variables used in standardising the raw data of the four catchments.

Catchment (area, km^2)	Variables	Rainfall	Flow
Silk Stream (31.25)	MINV	0.000	0.074
	MAXV	10.211	13.805
	MINT	-0.201	-0.201
	MAXT	20.626	27.885
Dollis Brook (23.99)	MINV	0.000	0.034
	MAXV	7.849	16.244
	MINT	-0.157	-0.291
	MAXT	15.855	32.813
Thrushel river (112.7)	MINV	0.000	0.175
	MAXV	9.500	123.656
	MINT	-0.190	-2.296
	MAXT	19.190	249.783
Dart river (248)	MINV	0.000	2.390
	MAXV	13.900	195.840
	MINT	-0.300	-1.510
	MAXT	30.100	395.570

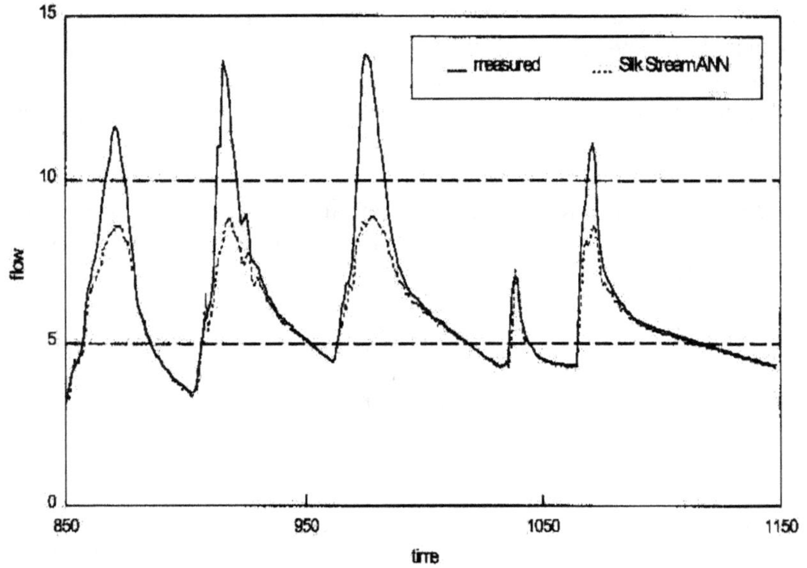

Figure 4-1 Extrapolation problem exists in the Silk Stream verification data set when the largest peak flow is not in the training data set (Adapted from Minns, 1998).

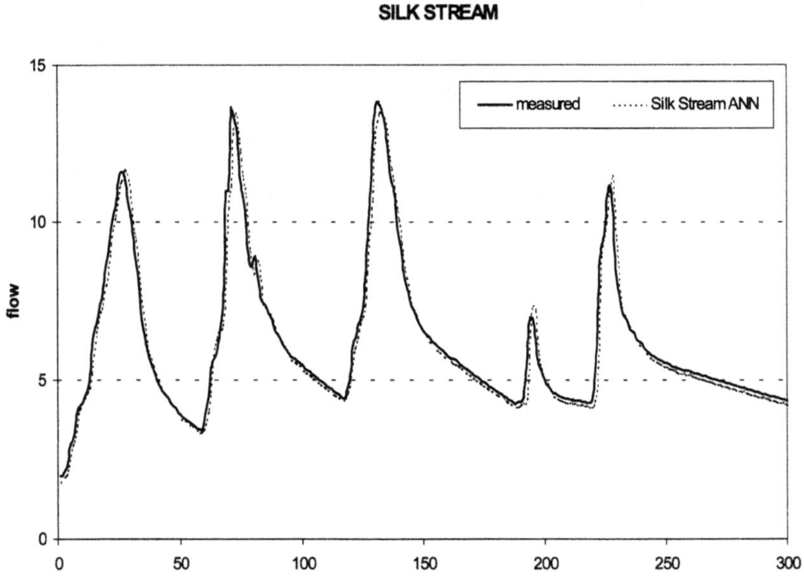

Figure 4-2 No extrapolation problem exists in the Silk Stream verification data set when applying the proposed method of standardising the inputs.

4.4.2 Dollis Brook catchment

Minns (1998) divided the Dollis Brook data into training and verification data sets, with the first 23 events being used for training, and the last 17 for verification. This division ensured that the maximum flow peak of 16.24 m^3/s was included in the training data and the maximum flow in the verification data of 3.55 m^3/s was smaller than the maximum in the training data set. In this case, a *CE* of 0.983 was obtained and there was no extrapolation problem.

To ensure that there was no side effect when the maximum value was in the training set, the same network was employed and the results were comparable with a *CE* of 0.980. To verify the new concept once more the training and verification sets were interchanged. The values of *MINV*, *MAXV*, *MINT* and *MAXT* for the rainfall and flow in this case are presented in Table 4.1. Applying the same procedure, the results from the network were obtained with the *CE* of 0.935. There was also no extrapolation problem in this case.

4.4.3 Thrushel river

Hettiarachchi et al (2003) divided the Thrushel river data into training, cross validation and testing data sets. Approximately 40% of the whole data set was used to train the network. An equal portion was kept for testing purposes and the rest were used in cross validation to prevent over-training the network. The network was constructed with five inputs (rainfall at time *t*, and the two previous time steps, and the flows at two previous time steps) and one output of flow at time *t*. There were three layers in the network with four nodes in the hidden layer. The characteristics of the three sets are summarised in Table 4.2.

Although the coefficient of correlation was 0.966, Hettiarachchi et al (2003) pointed out that the problem of extrapolation arose when a network was faced with extreme data that it had not seen before. Since the maximum flow in the testing set of 123.66 m^3/s is larger than those in the training and cross validation sets, these data were used to verify this new concept. The standardising variable values are shown in Table 4.1. The resulting coefficient of correlation is 0.982 and the coefficient of efficiency (*CE*) is 0.964. These two indices are based on different measures, the first being the square root of the coefficient of determination (CD) which differs from CE as shown in Table 3.3. In general the square of the correlation coefficient is only approximately equal to the CE. There was again no extrapolation problem in this case.

4.4.4 Dart river

Hettiarachchi et al (2003) divided the Dart river data into the same proportional sizes of sets as those of the Thrushel River and employed the same configuration of network. The characteristics of the two sets of streamflow data are summarised in Table 4.2.

This time, with the largest flow in the testing set, the coefficient of correlation was 0.975 and the problem of extrapolation still arose when a network was faced with extreme data that it had not seen before. Most of the calculated peak flows were almost the same value as the maximum flow in the training set.

To verify the new concept, the standardising variable values were adopted as given in Table 4.1. The results with the coefficient of correlation of 0.977 and the coefficient of efficiency (*CE*) of 0.954 were obtained. There was also no extrapolation problem in this case.

Table 4.2: Characteristics of the raw streamflow data.

Catchments	Statistics	Training Set	Cross Validation Set	Testing Set
Thrushel river	Minimum	0.616	1.642	0.179
	Maximum	61.46	75.427	123.656
	Average	7.034	11.527	8.558
	Std. Dev	8.484	11.992	12.842
Dart river	Minimum	3.797	8.419	2.386
	Maximum	137.088	165.311	195.839
	Average	31.436	44.227	31.032
	Std. Dev	29.270	38.535	37.307

Remarks: Std. Dev : Standard of deviation.

4.5 Case studies with including EMF in the raw data

Hettiarachchi et al (2003) suggested the used of an *Estimated Maximum Flood* (EMF) based on domain knowledge of the catchment, to train the ANN in order to avoid extrapolation problems. The EMF might be defined as the flood with a very small, but non-negligible, probability of annual exceedance. It was estimated by a procedure that is an extension of the unit hydrograph plus design storm approach in the UK Flood Studies Report (NERC, 1975). In this method, the unit hydrograph and the loss function were derived from a regional analysis of UK catchments, and the *Probable Maximum Precipitation* (PMP) storm profile is based upon the extrapolation of UK rainfall statistics. This type of regional domain knowledge was applied to the Thrushel and the Dart rivers, the relevant EMF's being 760 and 1,867 m^3/s, respectively. The authors then included the EMF hydrograph, together with the PMP storm profile, into the training data set.

In applying this EMF concept, several points can be made:

- The concept was to incorporate the biggest event in the training data set to guarantee that no observed events will ever be larger. Although there is a chance that new extremes are possible with climate change, in that case, a new EMF would need to be estimated again.

- The computation of PMPs and EMFs can be laborious, and by no means all countries have standardised procedures for this purpose.

Then these EMF data are tested with the new concept and the results are good without any extrapolation problem. To demonstrate that other activation function, for example, the hyperbolic tangent also works well, the EMF data are tested using this function and its standardised range. Since the EMF values are very large as compared to recorded data, the data have to be transformed into logarithms (Hettiarachchi et al, 2003). This transformation is needed, firstly, to prevent the extreme pattern (EMF) dominating the range of data and secondly, to distribute all patterns evenly in the network domain. All relevant standardising data and raw data are shown in Tables 4.3 and 4.4, respectively. The results of including the EMF values in the test data set for the Thrushel and Dart rivers are shown in Figures 4-3 and 4-4, respectively.

Table 4.3: Variables used in standardising the raw data for both rainfall and Log(Flow).

Catchment (km^2)	Variables	Rainfall	Log(Flow)
Thrushel river (112.7) CE = 0.964	MINV MAXV MINT MAXT	0.000 78.240 -1.565 79.805	-0.756 2.881 -0.821 2.946
Dart river (248) CE = 0.980	MINV MAXV MINT MAXT	0.000 74.200 -1.484 75.684	0.378 3.271 0.316 3.333

Table 4.4: Characteristics of the raw data (Log(Flow)).

Catchments	Statistics	Training Set	Cross Validation Set	Testing Set
Thrushel river	Minimum Maximum Average Std. Dev	-0.756 2.092 0.629 0.516	0.035 1.878 0.892 0.371	-0.210 2.881 0.666 0.423
Dart river	Minimum Maximum Average Std. Dev	0.579 2.213 1.347 0.396	0.925 2.218 1.503 0.352	0.378 3.271 1.350 0.513

Remarks: Std. Dev : Standard of deviation.

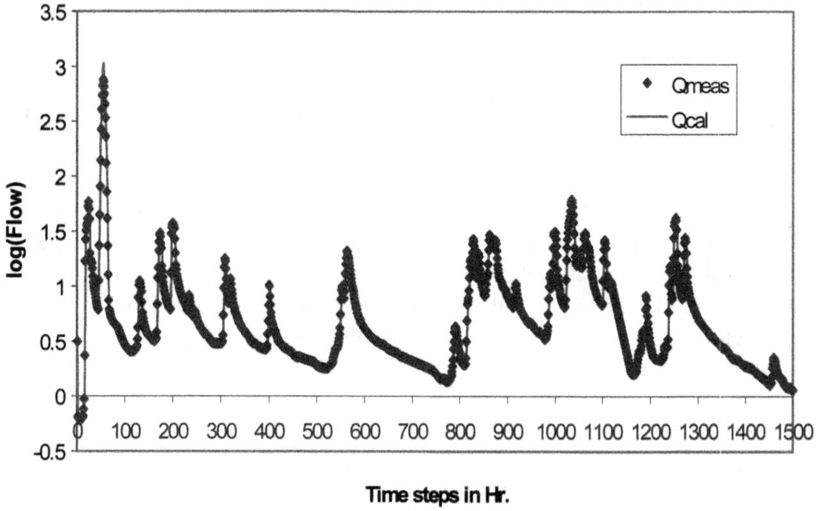

Figure 4-3 Thrushel river results with EMF in the test data set.

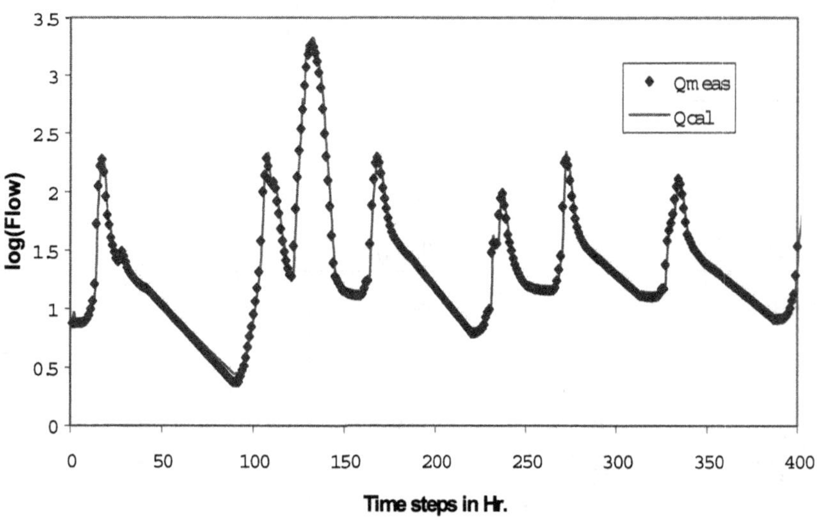

Figure 4-4 Dart river results with EMF in the test data set.

4.6 Summary

The new concept of standardisation has proven to be capable of managing the extrapolation problems previously encountered without any difficulty. The performance of ANNs was not affected at all when the maximum value was in the training set. For the purposes of illustration, the factor '*K2*' in equation (4.1), chosen to provide enough room to accommodate large events, was an arbitrary choice. However, even EMF values can also be simulated if logarithmic transformation is employed. Using the ungauged catchment procedures from the UK Flood Studies Report (NERC, 1975) and the tabulated annual flood statistics from the FREND Study (Gustard et al, 1989) the return periods associated with the *MAXT* values in Table 4.1 can be estimated. These values were found to range from about 6 years for the River Dart to almost 100 years for the River Thrushel. The North London catchments figures (using the FSR regional equations) were 15 years for Silk Stream and 33 years for Dollis Brook. Clearly, some thought is required in setting the value of the multiplier when applying the methodology to other catchments.

Since the relevant data used in this study are not sufficient for deriving such a return period of each event, a universally-applicable guideline for the factor '*K2*' cannot be developed at this stage. More details are given in Section 4.8.

4.7 Case studies with different activation functions

The question arises as to how the suggested standardisation procedures will perform with different activation functions. To answer this question, four different activation functions, the *linear*, *hyperbolic tangent*, *bipolar binary* and *sigmoid activation functions,* were employed to model the four catchments.

The four different activation functions were implemented for each catchment and the results are shown in Table 4.5, where Q_{PC} and Q_{PM} are calculated maximum value and measured maximum value respectively. Only the results for Silk stream are plotted in Figures 4-5 to 4-8, which represented the results of *linear*, *hyperbolic tangent*, *bipolar binary* and *sigmoid activation functions*, respectively.

The results shown in Table 4.5 reveal that the *Linear* activation function performs the fastest learning in all cases except the Dart river, which may have more non-linear characteristics than the others. The *hyperbolic tangent* activation function is the second fastest of all and the slowest is the *sigmoid* activation function. Based on CE and correlation coefficient measures, the performance of different activation functions is comparable.

The reason why both *linear* and *hyperbolic tangent* functions perform quicker than the rest is due to the back propagation algorithm, which depends on the derivative of the activation function being employed (equation 2.8 & 2.10). The *linear* and *hyperbolic tangent* activation functions share the same derivative value of one at their centroid, which are *two* and *four times* that of the *bipolar binary* function and the *sigmoid* function, respectively.

Table 4.5: Results of difference activation functions on the four catchments.

Catchments (area, km^2)	Variable	Activation Function				
		Linear	Hyper-Tangent	Bipolar Binary	Sigmoid	Hyper-Tangent & Linear
Dollis Brook (23.99 km^2)	Epoch	359	3,842	8,307	64,640	3,647
	RMSE	0.174	0.180	0.180	0.176	0.207
	CE.	98.339	98.364	98.378	98.374	97.736
	Q_{PC}/Q_{PM}	1.023	1.124	1.128	1.071	1.026
Silk Stream (31.25 km^2)	Epoch	3,225	17,943	17,943	205,836	15,001
	RMSE	0.338	0.374	0.374	0.366	0.458
	CE.	98.055	97.712	97.712	97.779	96.891
	Q_{PC}/Q_{PM}	1.070	1.122	1.122	1.114	1.105
Thrushel river (112.7 km^2)	Epoch	1,421	9,535	16,575	69,886	9,226
	RMSE	1.195	1.248	1.235	1.270	1.621
	CE.	99.135	99.011	99.031	98.985	98.773
	Q_{PC}/Q_{PM}	1.016	0.939	0.945	0.957	0.900
Dart river (248 km^2)	Epoch	4,357	1,272	1,993	9,961	1,372
	RMSE	8.002	8.192	7.932	7.826	7.435
	CE.	95.795	95.462	95.704	95.645	96.022
	Q_{PC}/Q_{PM}	1.113	1.136	1.140	1.090	1.118

Remarks: RMSE : Root mean square of error; CE : the coefficient of efficiency; Q_{PC} : calculated maximum value; Q_{PM} : measured maximum value.

The *linear* activation function learning speed outperforms the *hyperbolic tangent* since all its derivative values are constant and equal to one within the standardised range. Those of *hyperbolic tangent* vary between 0.84-1.00 so that the *linear* function is straightforward to apply as compared to the *non-linear hyperbolic tangent.*

According to the ratio of (Q_{PC}/Q_{PM}), the *linear* activation function tends to match the peak values best, while both *hyperbolic tangent* and *bipolar binary* functions, which belong to the same *hyperbolic* functions, tend to deviate from the peak most, as shown in Figure 4-5 to 4-8, (b).

Although the *linear* activation function may seem to perform best in many test catchments, it is not true for all cases since the ANN is a data driven model. Therefore, the results are also different from catchment to catchment. The *hyperbolic tangent* has the advantage of being non-linear and of moderate training speed. Therefore, it will mainly be employed in this study from now on.

A mixture of activation functions in which the linear activation function is employed in the output layer (Kaastra & Boyd, 1995 and Karunanithi et al, 1994) was also studied. As pointed out earlier, the performances of those non-linear activation functions are comparable but the *hyperbolic tangent* performs fastest among them. Therefore only the *hyperbolic tangent* will be mixed with the linear function to show that the suggested procedures still work well. Table 4.5 shows that, in general, the mixed activation functions perform very well but less accurately than those of the same non-linear activation functions in both hidden and output layers. There is only one exception in case of the Dart river, where the mixed activation functions perform better.

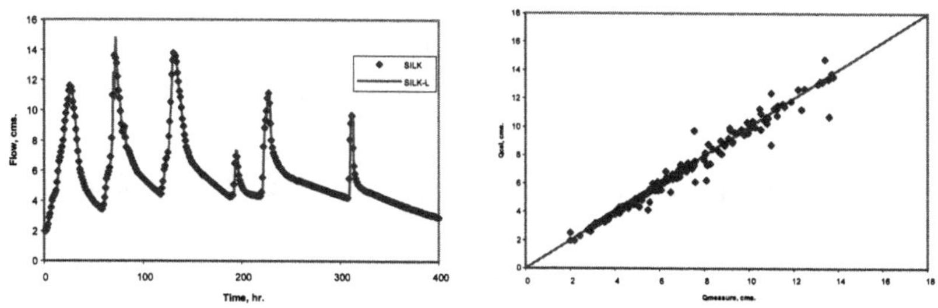

Figure 4-5 Linear activation function. (a) the time series for the testing data set (b) the scattergram.

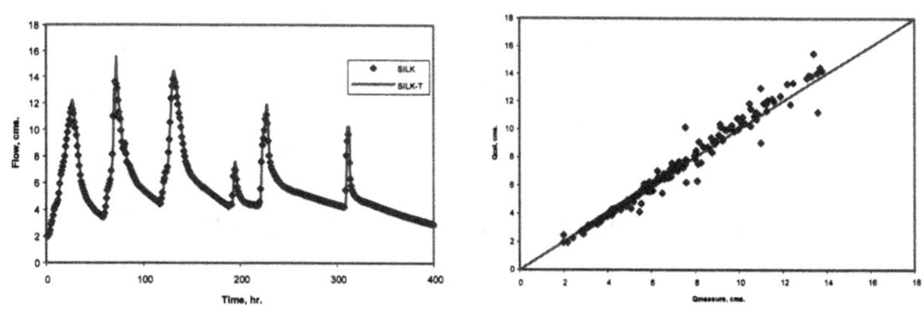

Figure 4-6 Hyperbolic tangent function. (a) the time series (b) the scattergram.

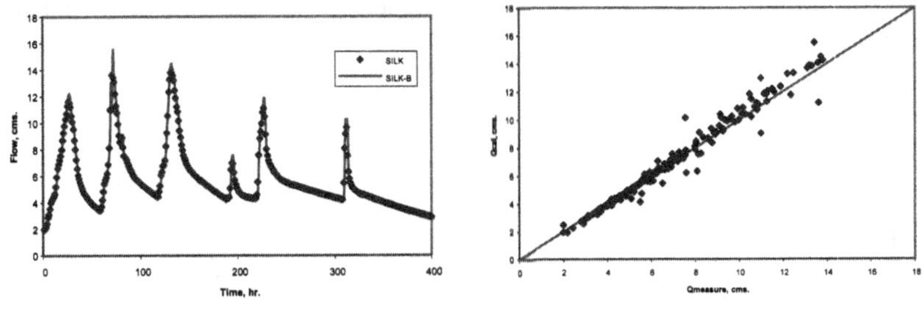

Figure 4-7 Bipolar binary function. (a) the time series for the testing data set (b) the scattergram.

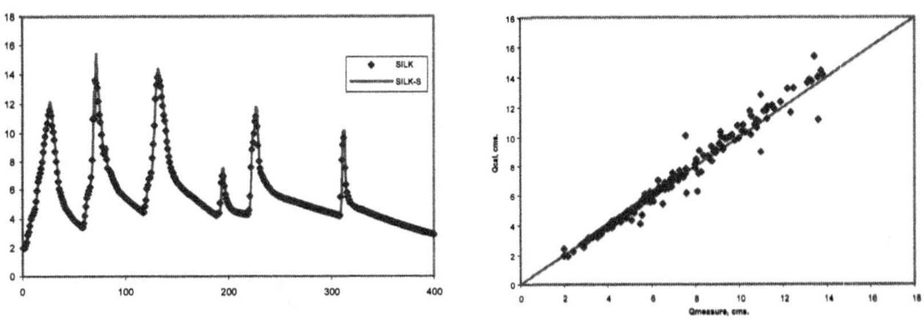

Figure 4-8 Sigmoid function. (a) the time series for the testing data set (b) the scattergram.

4.8 Different values of the factor *K2*

In order to evaluate the effect of employing values of the factor *K2* other than "*2*" as suggested in section 4.2, further tests were carried out on the Silk stream and the Dollis Brook. The characteristics of the Silk stream and the Dollis Brook are shown in Table 4.1. Varying the scale factor values from 1.5 to 20 as shown in Tables 4.6 and 4.7, incorporating hyperbolic activation functions for both hidden and output layers, the results can be summarised as follows:

The results of all various scale factor values are comparable. No distinct conclusion can be made as to which value is the best.

The larger the scale factor value, the more epochs are needed, the increase being exponential. As a result, the time consumed varies from minutes to hours on a Pentium III, 1.0 GHz personal computer.

In the case of logarithmic transformation, which is shown in the last rows of Table 4.6 and 4.7, the factor "2" is used, which, once transformed back to normal scale, equals 13.805 and 16.244 in case of the Silk stream and the Dollis Brook, respectively. The results are compared to the no-transformation cases and both are comparable in terms of coefficient of efficiency (CE). However the logarithmic transformation accelerated the learning speed by 28 and 10 times in case of the Silk stream and the Dollis Brook, respectively.

The factor "2" is appropriate in the sense that it provides a fast learning speed, accurate results and more room to accommodate future bigger flood magnitudes. If one expects to have flood magnitude approaching the EMF value, the logarithmic transformation is suggested, retaining the factor "2".

In all cases, the training will stop when comparisons with the test set failed to show significant improvements, therefore the number of epochs is not fixed and depends on those comparisons.

Table 4.6: Results of Silk stream with various scale factors.

Silk stream							
Scale	Epochs	RMSE	CE	$Qcal_{min}$	$Qcal_{max}$	Errmin	Errmax
1.5	8303	0.337	98.035	1.945	14.853	-2.623	2.445
2	12707	0.366	97.780	1.943	15.426	-2.437	2.567
3	22584	0.373	97.708	1.950	15.613	-2.424	2.555
4	38824	0.365	97.778	1.957	15.558	-2.476	2.507
5	55327	0.357	97.845	1.960	15.452	-2.531	2.461
6	80009	0.352	97.895	1.960	15.375	-2.578	2.422
7	95660	0.348	97.934	1.963	15.260	-2.615	2.390
8	128372	0.345	97.957	1.965	15.224	-2.640	2.371
10	197087	0.342	97.990	1.967	15.139	-2.681	2.337
15	421222	0.338	98.024	1.970	14.996	-2.741	2.290
20	684343	0.338	98.034	1.971	14.870	-2.775	2.259
13.805*	14676	0.028	98.480	1.929	13.715	-3.584	0.924

Remarks: RMSE : Root mean square of error; CE : the coefficient of efficiency; $Qcal_{min}$: calculated minimum value; $Qcal_{max}$: calculated maximum value; Errmin : minimum error value; Errmax : maximum error value; * logarithmic transformation.

Table 4.7: Results of Dollis Brook with various scale factors.

Dollis Brook							
Scale	Epochs	RMSE	CE	$Qcal_{min}$	$Qcal_{max}$	Errmin	Errmax
1.5	2822	0.1780	98.401	0.010	17.926	-4.540	2.697
2	3842	0.1802	98.364	0.011	18.260	-4.542	2.861
3	8141	0.1766	98.405	0.005	18.222	-4.535	2.709
4	10404	0.1762	98.390	0.017	17.893	-4.575	2.609
5	16692	0.1733	98.426	0.014	17.769	-4.561	2.451
6	22169	0.1731	98.419	0.014	17.590	-4.576	2.365
7	30934	0.1718	98.434	0.011	17.516	-4.567	2.275
8	36041	0.1732	98.397	0.021	17.335	-4.591	2.243
10	53925	0.1736	98.381	0.022	17.178	-4.601	2.162
15	109722	0.1759	98.320	0.026	16.903	-4.627	2.060
20	199475	0.1751	98.327	0.026	16.838	-4.621	1.986
16.244*	10849	0.0491	99.030	0.015	17.139	-3.992	3.542

Remarks: RMSE : Root mean square of error; CE : the coefficient of efficiency; $Qcal_{min}$: calculated minimum value; $Qcal_{max}$: calculated maximum value; Errmin : minimum error value; Errmax : maximum error value; * logarithmic transformation.

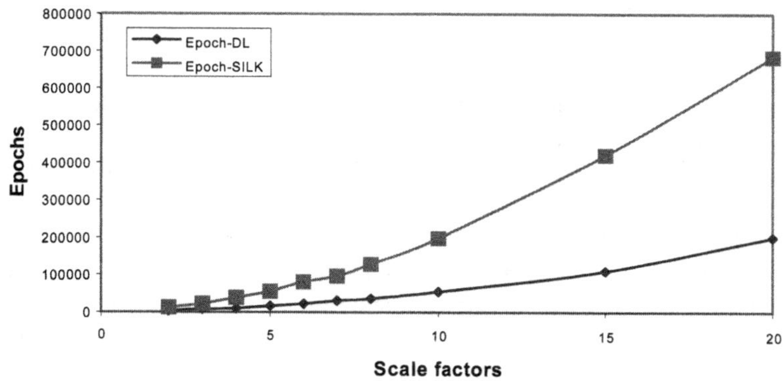

Figure 4-9 Effect of various scale factors on number of Epochs.

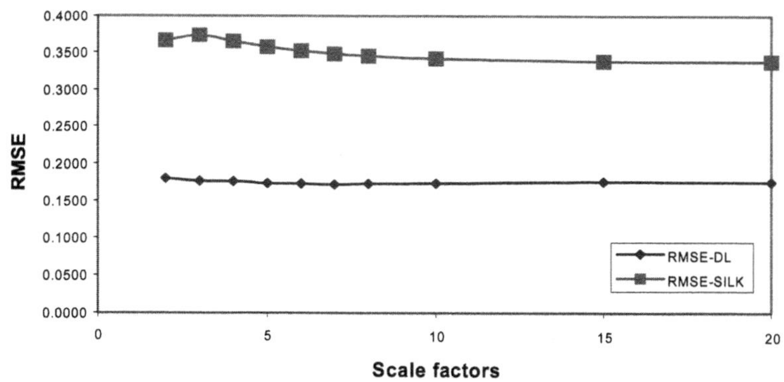

Figure 4-10 Effect of various scale factors on Root Mean Square Error.

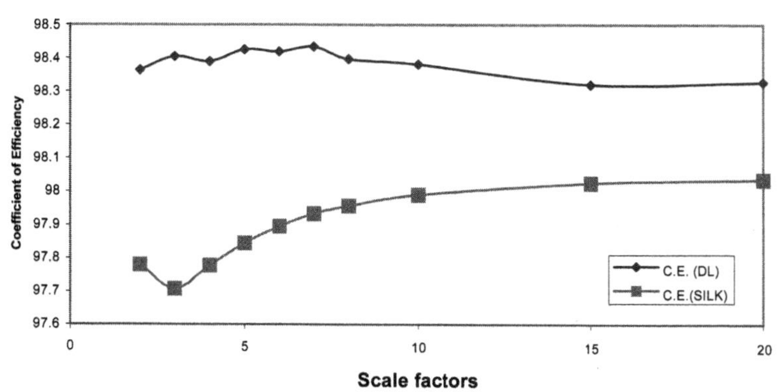

Figure 4-11 Effect of various scale factors on Root Mean Square Error.

4.9 Summary

The introduced new concept of standardisation has been tested and found to work well in all cases no matter what maximum value is included or excluded in the training data set. It was proved with a Sigmoid function and then extended to cover various activation function eg. Bipolar Binary function, Linear function, etc.

The concept has proven to work well even with the extreme values like the EMF included in the data set. This concept employs the factor 'K2' to provide space to accommodate the big events to fit in and the value of '2' is recommended. Various values of K2 have been studied with no significant improvement in results except the point that the number of epochs is increased with a higher value of K2. Further work is necessary to establish whether K2 could be related in some way to a low-frequency quantile of the flood frequency distribution for a given catchment.

5

Recurrent Neural Networks

5.1 Introduction

As demonstrated in Chapter 3, when FFMLPs are employed to forecast flows at longer time horizons, the results tend to degrade rapidly in a nonlinear fashion. This degradation can be attributed to several sources, including phase error, propagation of previous error in time, inadequate information provided to the network, etc. It is believed that *Dynamically-Driven Recurrent Networks* can provide the possibility to forecast many steps ahead with less difficulty. Therefore, the two *Recurrent Networks,* namely the *Elman Recurrent Networks* and the *Jordan Recurrent Networks*, were studied. To avoid the extrapolation problems, the scheme introduced in Chapter 4 was implemented throughout these studies. Although there are a number of commercial software packages dealing with neural networks, none of them gives away the source code. Hence, it is almost impossible to modify the program to take into account the extrapolation problem. Therefore, a neural network program was developed by the author first in C++ and later in Fortran languages and used throughout this study.

There are two ways to march forward in time. The first one is to use the same set of input data at time t to predict either single-step-ahead (e.g. at time $t+1$) or n-step-ahead (e.g. at time $t+n$). The other is the *time marching scheme* that preserves the same input structure to predict one-step ahead, but repeated consecutively.

The two approaches are implemented using the feed forward network. More details are given in Section 5.2 with two example cases. The River Brue is the first example in which the same data at time t are applied to predict n-step-ahead (e.g. at time $t+n$). This example serves as a basis for comparison with the results from the two recurrent networks. The River Thrushel is the second example to which the time marching scheme is applied.

The advantage of using the Elman and the Jordan recurrent networks over the feed forward network will be discussed through the example of the River Brue in Sections 5.3 and 5.4, respectively. Section 5.5 elaborates more on the Jordan recurrent network with recurred error. Although the future error cannot be known before the measured data are collected, this concept of recurred error leads to the forward shift operator that will be given in Chapter 6. Finally, concluding remarks are given at the end of this Chapter.

5.2 Feed forward network

Three examples are given in this Section. The River Brue is studied using only the flow data as the input to the network. For example, the inputs for predicting n-time-step-ahead (QP_{t+n}) are composed of three inputs: one concurrent and two antecedent flows (Q_t,Q_{t-1},Q_{t-2}). In all cases throughout these studies, the hidden layer consisted of three nodes and the output layer consisted of only one output node. The reason for selecting three hidden nodes was to avoid over-parameterisation problems and to show that fewer hidden nodes can give accurate results. Then, the time marching scheme is applied to data from the Rivers Thrushel and Chumporn. The Thrushel (in Section 5.2.2) is studied using both rainfall and flow data as the input to the network, while the Chumporn example (in Section 5.2.3) employed upstream water level data as input.

5.2.1 Case I: the Brue River

The Brue catchment data (see Appendix I) were resampled into 30-minute-interval according to Sugawara's criterion (see Chapter 3 for more details). The architectures of FFNs for predicting n-time-step-ahead (QP_{t+n}) were 3-3-1. There were three inputs in the input layer comprising one concurrent and two antecedent flows (Q_t,Q_{t-1},Q_{t-2}). The results obtained are summarised in Table 5.1, and reveal that the input did not provide enough information for the network at longer forecast times. Therefore, the coefficient of efficiency drops rapidly when proceeding forward in time.

Table 5.1: Feed Forward schemes on River Brue data with 30 minute time intervals.

Forecast Time (hr)	Epochs	CE	RMSE	MAE
0.50	7438	99.89	0.57	0.56
1.00	7501	99.59	0.45	0.42
2.00	7284	97.53	0.58	0.40
3.00	7002	91.25	0.94	0.42
4.00	7953	86.56	1.19	0.57
6.00	6332	68.66	1.73	0.57
12.00	6229	44.81	2.27	0.78

Remarks: CE : the coefficient of efficiency; RMSE : Root mean square of error; MAE : Mean absolute error.

The Brue catchment was the subject of a separate study employing a physically-based model (SHETRAN) and a semi-distributed conceptual model (ARNO), the results of which were summarised by Mellor et al (2000). For one step ahead forecasts on a testing data set, SHETRAN produced a CE-value of 0.89 and the ARNO model gave a value of 0.80. The ANN is seen from Table 5.1 to outperform both models up to *three* hours ahead.

5.2.2 Case II: the Thrushel River

The concept of the time marching scheme is to keep the lag structure among the input data constant. In the case of the Thrushel River, the inputs for predicting one time step ahead (QP_{t+1}) were composed of six inputs: the concurrent rainfall and two antecedent rainfalls (RF_t, RF_{t-1}, RF_{t-2}), one concurrent and two antecedent flows (Q_t,Q_{t-1},Q_{t-2}). The time step used was one hour. Therefore, to march in time, the six inputs for the second time step ahead (QP_{t+2}) are: the concurrent rainfall and two antecedent rainfalls (RF_t, RF_{t-1}, RF_{t-2}) since no future rainfall could be incorporated, one concurrent and one antecedent flow (Q_t,Q_{t-1}) plus one predicted flow (QP_{t+1}). The input of predicted flow (QP_{t+1}) helps provide useful information to the network. Finally, the six inputs for the third time step ahead (QP_{t+3}) are again the concurrent rainfall and two antecedent rainfalls (RF_t, RF_{t-1}, RF_{t-2}), and the concurrent flow (Q_t) plus two predicted flows (QP_{t+1}, QP_{t+2}), etc. The architectures of FFNs for predicting n-time-step-ahead (QP_{t+n}) in this case were 6-3-1.

The results are shown in Figures 5-1 to 5-3. The results using the standard feed-forward networks (FFNs), denoted as TF1, TF2 and TF3 for one, two and three step ahead, respectively, are shown in Table 5.2. The first two results are very good with the coefficients of efficiency between the calculated and measured data being greater than 95%, while the last one is satisfactory (89.5%) but less accurate, since available rainfall could not provide further information to the network. The RMSEs are increased from 1.87 to 3.93 m^3/s when forecasting forward in time. If the CEs are considered as a measure, all results would be very good in the view of at least one author (Shamseldin, 1997) since they are higher than about 90 %.

Table 5.2: Testing Results of Feed Forward schemes on the Thrushel river data of hourly interval.

Variables	TF1	TF2	TF3
EPOCHS:	8295	13826	11028
CE (%)	97.64	95.29	89.47
RMSE :	1.87	2.63	3.93
MAE :	1.08	1.14	1.65

Remarks: CE : the coefficient of efficiency; RMSE : Root mean square of error; MAE : Mean absolute error; TF1, TF2 & TF3: Thrushel Feed Forward predicted flows at 1, 2 & 3 step ahead,respectively.

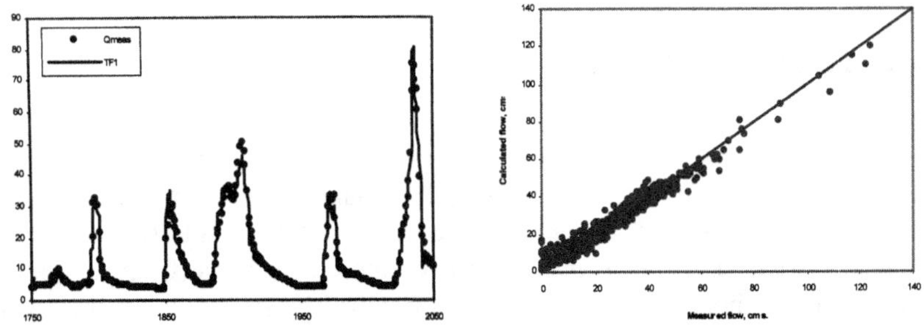

Figure 5-1 FFN for 1 time step ahead (TF1). (a) Time series (b) Scattergram.

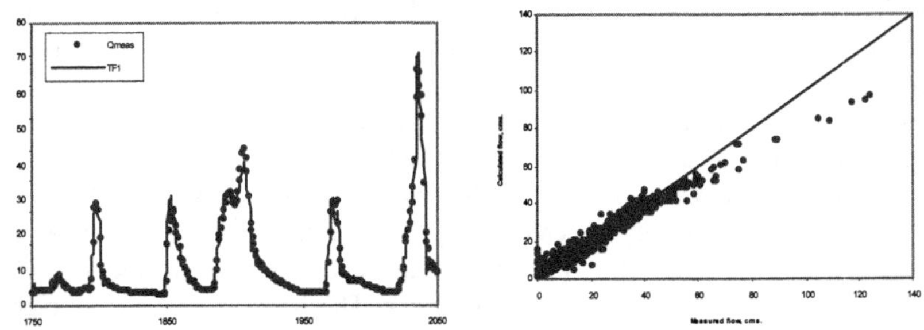

Figure 5-2 FFN for 2 time step ahead (TF2). (a) Time series (b) Scattergram.

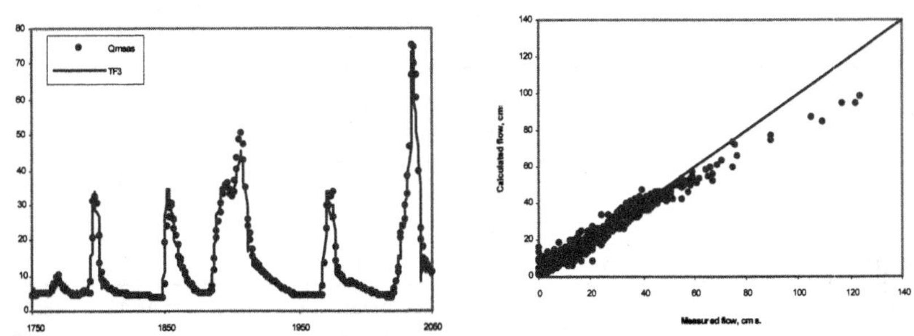

Figure 5-3 FFN for 3 time step ahead (TF3). (a) Time series (b) Scattergram.

Although the time marching scheme has been implemented in this case, the results show large deviations in peaks at longer forecast time horizons greater than three hour ahead. These are due to the rainfall data being unable to provide further information to the network and the subsequent errors propagating through time.

The results here and the results of the previous section indicate that, without incorporating additional features into a FFN, predictions for both the Brue and the Thrushel catchments deteriorate significantly for prediction horizons exceeding three hours irrespective of whether using n-time-step-ahead or time marching.

5.2.3 Case III: the Chumporn River

Somsin & Varoonchotikul (2002) applied the FFMLP with a time marching scheme to the Chumporn river in Thailand (see Appendix I), for which the data are available at hourly time intervals. There are three gauging stations, one at the outlet named X.158 and the other two in the headwaters of the catchment named X.46 and X.64, respectively. In this study, it was observed that the upstream flow data are more useful than the rainfall data since they are the measured outcomes of rainfall over the sub-catchments.

The inputs for predicting the water level at one time step ahead at the outlet ($PX158_{t+1}$) are derived using correlation analysis. They are composed of seven inputs: two antecedent (lag 6 and 7 hours) water levels at gauging sites X46 ($X46_{t-6}$,$X46_{t-7}$) and X64 ($X64_{t-6}$,$X64_{t-7}$) and three antecedent water levels at station X158 ($X158_{t}$, $X158_{t-1}$, $X158_{t-2}$).

Therefore, to march in time, the seven inputs for the second time step ahead ($PX158_{t+2}$) are: the two antecedent (lag 5 and 6 hours) water levels at X46 ($X46_{t-5}$,$X46_{t-6}$) and X64 ($X64_{t-5}$,$X64_{t-6}$), the concurrent and one antecedent of water levels at X158 ($X158_{t}$, $X158_{t-1}$) and one predicted water level ($PX158_{t+1}$). The input of predicted $PX158_{t+1}$ helps provide useful information to the network. This procedure was repeated up to a six-hour forecast.

The forecast results of one time step ahead obtained with the model are shown in Figure 5-4 for the period of testing and the results for six-hour ahead forecasting are shown in Figure 5-5. The X158meas, PX158-1 and DelWL are measured water level, calculated water level and the difference of calculated water level from the measured water level, respectively. The results show good agreement between the calculated and measured river stage at X.158. The coefficient of efficiency is found to be 99.66 %. The maximum error in the prediction for the whole period is less than 0.50 metre and the average value of error is about 0.03 metre with standard deviation of 0.07 metre. Figure 5-4 reveals also that there is no phase error in the results.

The calculated peak water levels fluctuate around 7.00 m and decrease through time as can be seen from Table 5.3. The worse case was five-hour forecasting, which underestimated the peak and missed the true value by 0.26 m. The selected results of six-hour ahead are shown in Figure 5-5, which reveals that the phase errors are still propagated in time but are limited to some extent.

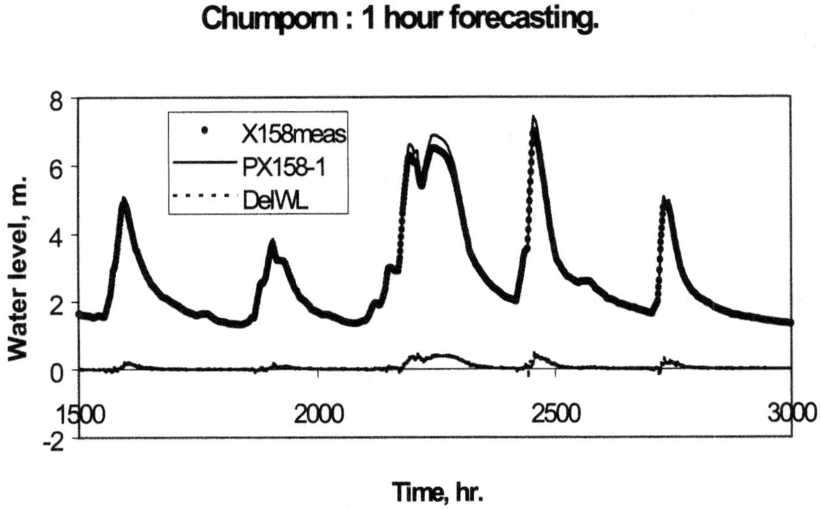

Figure 5-4 Time series of measured and calculated water level at X.158 for one hour ahead forecasting. (X158meas: measured water level; PX158-1: calculated water level; DelWL: difference of calculated water level)

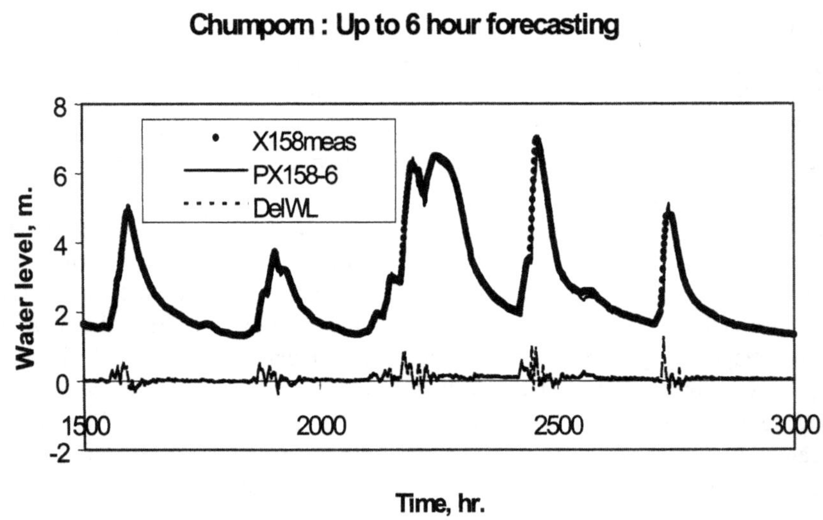

Figure 5-5 Time series of measured and calculated water level at X.158 for six-hour ahead forecasting. (X158meas: measured water level; PX158-6: calculated water level; DelWL: difference of calculated water level)

Table 5.3: Results of up to six-hour forecasting.

Forecasted	1-hour	2-hour	3-hour	4-hour	5-hour	6-hour
CE (%)	99.66	99.60	99.15	98.82	98.24	99.08
Cal WL, m	7.43	7.12	7.49	7.57	6.74	7.07
Min Error, m	-0.24	-0.52	-0.81	-1.02	-1.23	-1.25
Max Error, m	0.47	0.38	0.69	0.81	0.60	0.77

Remark: Cal WL : calculated water level in meter, the maximum measured water level is 7.00 meter; Min Error : minimum error value; Max Error : maximum error value.

5.3 Elman recurrent network

The *Elman* recurrent network (ERN) is characterised by recurring its own output signal from the hidden layer back to the context unit in the input layer (see Chapter 2 for more details). Dibike et al (1999) had studied this scheme and applied it to the Apure River in Venezuela. By comparing the results with those of FFMLPs, those authors concluded that both networks were comparable and no advantage could be gained with the Elman scheme. Furthermore, the ERN required more training time to obtain convergence. The authors suggested that this was due to the imposition of an additional context pattern on the input layer which increased the number of weights that had to be adjusted during training.

The Brue catchment data were employed to re-evaluate this conclusion. The same input data used in Section 5.2.1 are employed in this section. The architectures of ERNs for predicting n-time-step-ahead (QP_{t+n}) were 6-3-1. There were three exogenous inputs to the input layer comprising one concurrent and two antecedent flows (Q_t,Q_{t-1},Q_{t-2}), plus three recurred outputs from the hidden layer in the context unit. The results obtained are plotted in Figure 5-6 to 5-9 and shown in Table 5.4.

Comparison of Table 5.1 and Table 5.4 reveals that the extra information from the context unit did not provide enough information for the network to gain more accuracy than the FFMLP. Since the context units always provide lagged information, the errors are inherent in the forecast values, and are therefore not helpful in managing the phase errors. Figure 5-6 and 5-7 confirm the results of Dibike et al (1999) that the ERN scheme offers no advantages over the FFMLP and the training times of ERN schemes are higher than those of FFMLPs.

In terms of the Root Mean Square Error (RMSE), the Elman scheme does better than the FF network, when forecasts less than two hours (4 time intervals) ahead are made, and was outperformed for the longer forecasting horizons.

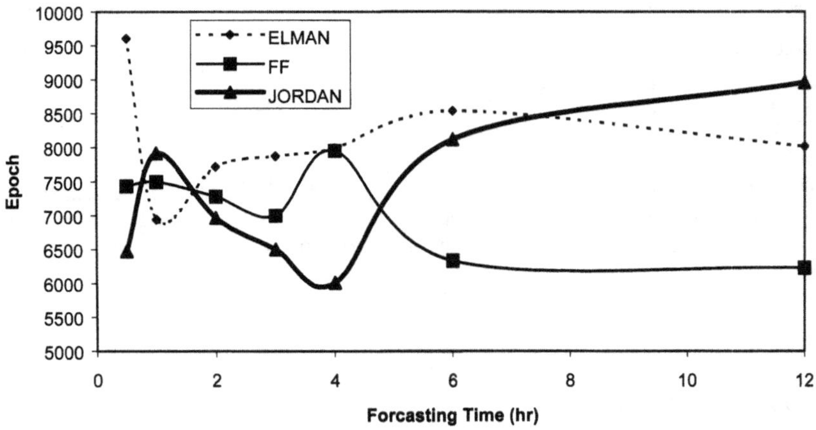

Figure 5-6 Epoch required by Elman, Jordan and FF networks

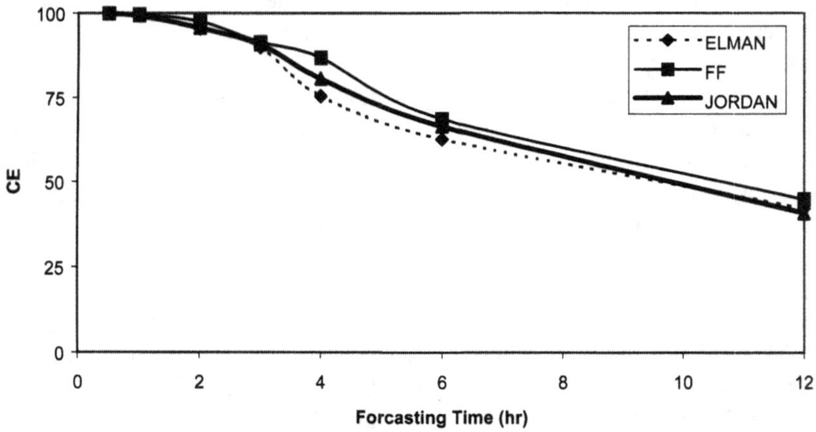

Figure 5-7 Coefficient of efficiency (CE) of Elman, Jordan and FF networks

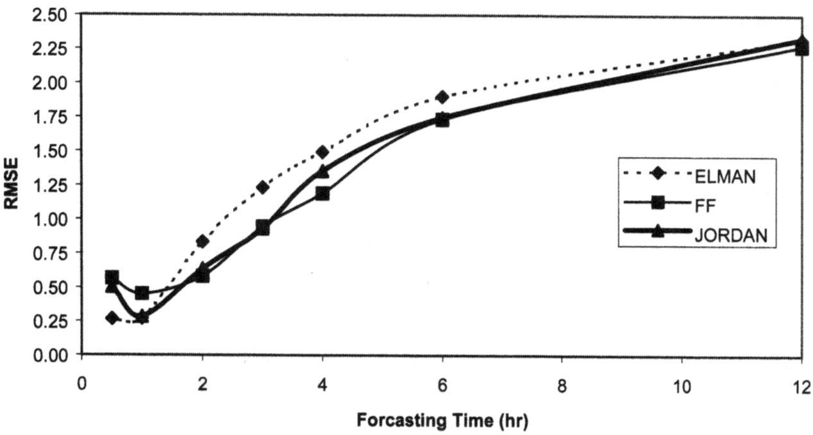

Figure 5-8 Root Mean Square of Error (RMSE) of Elman, Jordan and FF networks

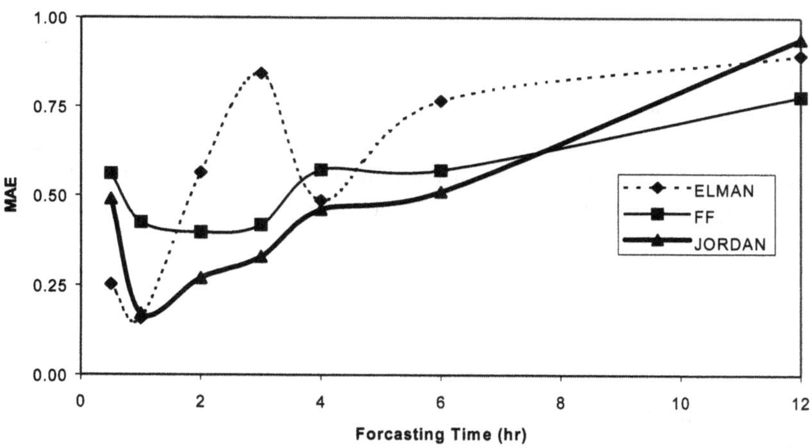

Figure 5-9 Mean Absolute Error (MAE) of Elman, Jordan and FF networks

Table 5.4: Elman Recurrent schemes on River Brue data with 30 minute time intervals.

Forecast Time (hr)	Epochs	CE	RMSE	MAE
0.50	9603	99.89	0.27	0.25
1.00	6943	99.27	0.27	0.16
2.00	7720	95.25	0.83	0.56
3.00	7876	89.71	1.23	0.84
4.00	7993	75.34	1.49	0.49
6.00	8537	62.52	1.90	0.77
12.00	8015	42.18	2.32	0.89

Remarks: CE : the coefficient of efficiency; RMSE : Root mean square of error; MAE : Mean absolute error.

5.4 Jordan recurrent network

The *Jordan* recurrent network (JRN) differs from *Elman* recurent network (ERN) by recurring its output signal from the output layer back to the context unit in the input layer (see Chapter 2 for more details). This context unit provides useful information to the network, since the lag flow data generally has a stronger auto-correlation than the rainfall data. Therefore, good results can be obtained if one provides a window of lagged flow data to the network instead of rainfall (Hall & Minns, 1993).

Two examples are given in this Section, the Rivers Brue and Thrushel. The River Brue employed only the flow data as in the previous cases of the FFN and ERN for comparison purpose. Then the Thrushel is studied using both rainfall and flow data as input.

5.4.1 Case I: the Brue River

The same data for the Brue catchment were employed with the context units of recurred outputs. The architectures of JRNs for predicting n-time-step-ahead (QP_{t+n}) were 6-3-1. There were three exogenous inputs in input layer comprising one concurrent and two antecedent flows (Q_t,Q_{t-1},Q_{t-2}), plus three recurred outputs (QP_{t+n-1},QP_{t+n-2},QP_{t+n-3}) from the output layer in the context unit. The results were compared with the ERN scheme and shown in Table 5.5 and Figure 5-6 to 5-9. Figure 5-7 reveals that the JRN performed better than the ERN in terms of the Coefficient of Efficiency and the required training times are generally less than those of the ERN. Figure 5-8 and 5-9 show that both RMSE and MAE of JRNs are smaller than that of the ERNs. Therefore, Jordan recurrent networks will be studied in greater depth.

More importantly, comparison of Table 5-5 and Table 5-1 shows very little difference in performance between the standard FFN and the recurrent ERN and JRN for n-time-step-ahead prediction.

Table 5.5: Comparison of Elman and Jordan Recurrent schemes on River Brue data with 30 minute time intervals.

Forecast Time (hr)	Epochs		CE		RMSE		MAE	
	ERN	JRN	ERN	JRN	ERN	JRN	ERN	JRN
0.50	9603	6473	99.89	99.84	0.27	0.50	0.25	0.49
1.00	6943	7915	99.27	99.29	0.27	0.28	0.16	0.17
2.00	7720	6969	95.25	95.52	0.83	0.64	0.56	0.27
3.00	7876	6501	89.71	90.48	1.23	0.93	0.84	0.33
4.00	7993	6014	75.34	80.51	1.49	1.35	0.49	0.46
6.00	8537	8118	62.52	66.42	1.90	1.75	0.77	0.51
12.00	8015	8957	42.18	40.78	2.32	2.33	0.89	0.94

Remarks: CE : the coefficient of efficiency; RMSE : Root mean square of error; MAE : Mean absolute error.

5.4.2 Case II: the Thrushel River

Varoonchotikul et al (2002b) applied the JRN with a time marching scheme to the Thrushel river (see Appendix I), for which hourly rainfall and flow data were available. The inputs for predicting n-time-step ahead (QP_{t+n}) are composed of seven inputs: the concurrent rainfall and two antecedent rainfalls (RF_t, RF_{t-1}, RF_{t-2}), one concurrent flow (Q_t) plus three recurred outputs (QP_{t+n-1},QP_{t+n-2},QP_{t+n-3}) in the context unit. Therefore, the concurrent flow (Q_t), concurrent rainfall and two antecedent rainfalls (RF_t, RF_{t-1}, RF_{t-2}) are kept constant when marching forward in time since no future flow and rainfall could be incorporated. Only three inputs in the context unit will be changed dynamically. The architectures of JRNs for predicting (QP_{t+n}) using time marching were 7-3-1.

The results using Jordan Recurrent networks, shown in Figure 5-10 and 5-11, are denoted in Table 5.6 as JTF2 and JTF3 for two and three steps ahead, respectively. These results are comparable to those of the feed forward network in Table 5-2, and allow the following comparisons to be made. Firstly from Table 5.6, the JRN requires about 40% fewer epochs to obtain the same results. Secondly from scattergram of Figures 5-2, 5-3, 5-10 and 5-11, the observed peaks were underestimated more by the FFN than the JRN, eventhough the extrapolation problem has been controlled by the scheme suggested in Chapter 4. The reason for this underestimation is that the supplied rainfalls in the FFN are not sufficient to suppress the phase errors, which propagate in time. On the other hand, the recurred output in the context unit provides better information than the lag exogenous output used in the FFN. One possible method for treating these phase errors will be introduced in Chapter 6.

Table 5.6: Testing Results of Jordan Recurrent schemes on the Thrushel river data of hourly interval.

Variables	JTF2	JTF3
EPOCHS:	6709	6171
CE (%)	95.67	89.16
RMSE :	2.52	4.26
MAE :	1.10	2.47

Remarks: CE : the coefficient of efficiency; RMSE : Root mean square of error; MAE : Mean absolute error; JTF2 & JTF3: Thrushel Jordan Recurrent predicted flows at 2 & 3 step ahead,respectively.

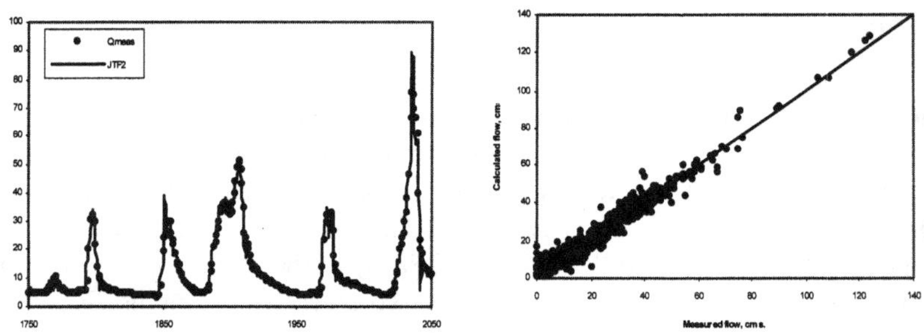

Figure 5-10 JRN for 2 time step ahead (JTF2). (a) Time series (b) Scattergram.

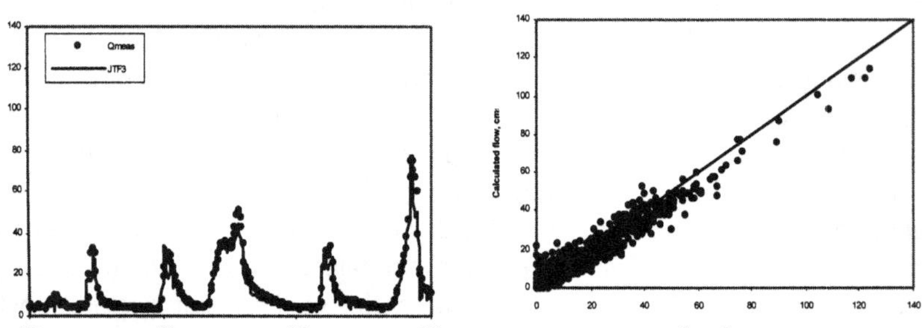

Figure 5-11 JRN for 3 time step ahead (JTF3). (a) Time series (b) Scattergram.

5.5 Jordan recurrent network with Future Errors

The advantage of a JRN is the possibility to recur the *outputs* and the *errors* back to the context units at the end of calculation. If it were possible to recur the errors, which are the differences between the computed and the measured values, back to the context units, the accuracy of the forecasting values in excess of two hours can be dramatically improved, as shown in Table 5.7, for the River Brue one step ahead forecasting.

Table 5.7: Comparison of Jordan Recurrent schemes without and with recurred Errors in the context units on River Brue data with 30 minute time intervals.

Forecast Time (hr)	Epochs		CE		RMSE		MAE	
	W/o error	With error	W/o error	With error	W/o error	With error	W/o error	With error
0.50	6473	7098	99.84	99.89	0.50	0.18	3.41	3.39
1.00	7915	6616	99.29	99.87	0.28	0.14	6.62	2.91
2.00	6969	8716	95.52	99.90	0.64	0.21	10.38	3.16
3.00	6501	10070	90.48	99.90	0.93	0.11	14.75	3.00
4.00	6014	15014	80.51	99.87	1.35	0.11	22.64	3.46
6.00	8118	5808	66.42	99.88	1.75	0.16	32.47	2.91
12.00	8957	6173	40.78	99.91	2.33	0.56	40.68	3.41

Remarks: CE : the coefficient of efficiency; RMSE : Root mean square of error; MAE : Mean absolute error.

The architectures of JRNs without recurred errors for predicting n-time-step-ahead (QP_{t+n}) were 4-3-1. The input comprised of one concurrent flow (Q_t) plus three recurred outputs (QP_{t+n-1},QP_{t+n-2},QP_{t+n-3}) from the output layer in the context unit. In case of JRNs with recurred future errors for predicting n-time-step-ahead (QP_{t+n}), the architectures were 5-3-1. The extra input was the recurred future error in the context unit.

In all cases, the coefficients of efficiency are higher than 99.80%, which are clearly better than either the ERN or JRN scheme without recurred errors. The number of epochs is high in the 4-hour ahead forecasting, moderate for 2 and 3 hours ahead, and about the same for the rest.

By recurring the errors back to the context unit in the JRN scheme, the network gains extra information about the behaviour of the river basin, which results in higher coefficients of efficiency. This dynamic behaviour is captured far better than when the ERN scheme is applied.

In the case of FFMLPs, no calculation is carried out since FFMLPs will be JRNs if one recurs right away the calculated output to the input layer. Even if one uses antecedent streamflow as input, the network will perform worse than the JRN because the longer the forecasting time horizon, the weaker the auto-correlation and the poorer the coefficient of efficiency.

The recurred errors work as a sort of *forward shift operator* to the inputs, which in this case are the *recurred outputs*. Therefore, the key to success in managing the *phase errors* is to incorporate the information, which encapsulates these errors. This problem will be dealt with further in Chapter 6.

5.6 Concluding Remarks

There are two possibilities to predict future outcomes. The first scheme employs the same set of input data for all the forecast horizons and is referred to here simply as n-time-step-ahead prediction. The second scheme maintains the structure of input, which is then applied repetitively until the appropriate forecast horizon is reached. This scheme is referred to as time marching. It is found that the latter can inhibit some phase error much better than the first one (Varoonchotikul, 2002b and Somsin & Varoonchotikul, 2002). Therefore, the time marching scheme was adopted throughout the rest of this study.

Different recurrent networks have been introduced. The JRN has been shown to capture more information than the ERN. The reason is that the recurred outputs from the output layer provide more useful information than the outputs from the hidden layer.

When applying the JRN with a time marching scheme to forecast floods using rainfall runoff models, the results are much better than the FFMLP equipped with the exogenous recurred outputs. By comparing the scattergrams of the results of both networks, the JRN can capture the peak values better than the FFMLP. The reason is that the JRN exhibits a dynamic behaviour, which can in parallel adjust its weights and outputs, while the FFMLP behaviour is static.

The time marching scheme introduced here, when applied to a flood routing model, can suppress or limit some phase errors that otherwise would propagate and be amplified in time. In most cases, these phase errors degrade the forecasting results very rapidly. Although the networks can be used to predict floods quite well, it depends very much on the input data.

The Feed Forward Networks (FFNs) are straightforward to use and higher learning rates can be used, while the Jordan Recurrent Networks are more complex in that their learning rates must be carefully selected to provide stability to the network (Varoonchotikul et al, 2002b). The drawback of selecting this learning rate is compensated by cutting down the epochs by 40% as compared to an FFN.

6
Choice of input

6.1 Introduction

A primary goal of any rainfall-runoff modelling exercise is the prediction of a variable in time and space from a given set of rainfall inputs. In general, distributed rainfall input can provide more information than the areal average rainfall but it also poses difficulties in data handling, processing time, etc. In this chapter, the choice of either an areal average rainfall or a set of distributed rainfall will be studied in detail.

As pointed out in previous chapters, when forecasting several time intervals ahead, phase errors may propagate and diminish the accuracy of the forecasts. A *forward shift operator* of some description is therefore required to improve the performance of forecasting schemes. To some extent, the JRN architecture provides a memory whose operation may be viewed in a feed-forward sense. However, an alternate approach to controlling the phase error may be found in the provision of an initial forecast using a conventional approach, with the ANN incorporated as an error correction algorithm. This modelling strategy was therefore investigated using the *First Difference Transfer Function* (FDTF) method of Duband et al (1993), which is essentially a unit hydrograph based approach, to provide the initial forecast. The results obtained are illustrated by examples from two catchments in Section 6.4.

6.2 Average rainfall or Distributed rainfall

In general the information obtained from the distributed rainfall stations are more valuable than those of the average basin rainfall, when distributed physically-based models, e.g. 'SHE', are employed. For Neural Network based models, whether the distributed rainfall will provide more advantages than the average basin rainfall or not, remains to be demonstrated. To investigate this matter, the River Brue (see Appendix I) was selected since there are 49 automatic-rainfall-gauging stations in the basin and the basin average rainfall is derived from these stations. To simulate rainfall-runoff relationship by employing these stations would require enormous memory resources and calculation time (Maier & Dandy, 2000). Furthermore, the program run by the author could not accommodate all these data due to limiting memory (RAM). Therefore, five representative sub-basin rainfalls were carefully selected and used instead of basing the calculations on all stations. These stations are at Alford Fields, Kite Lane, Crabtree Lane Field, Hadspen House Fields and Cooks Farm Field. Examples of the basin average and four of these representative stations distribution are shown in Figure 6-1 and 6-2, which demonstrate the variability of rainfall distribution over the basin during the period of October 17, 1994 to November 1, 1994. The rainfall patterns are similar but vary in magnitudes or depths which depend on how close are the position of the gauges to the centre of the storms. The five gauges appear to represent the basin average well for this particular catchment.

The Feed Forward Network (FFN) is selected in this study to compare both sets of data. In the case of the average basin rainfall, the FFN is composed of three layers. The configurations of the FFN models were determined based on correlation analysis and past experience. The architectures of FFNs for predicting one time step ahead (Q_t) were 6-3-1. There were six inputs in input layer comprising three antecedent rainfalls (RF_{t-6}, RF_{t-7}, RF_{t-8}), and three antecedent flows ($Q_{t-1}, Q_{t-2}, Q_{t-3}$). In all cases, the hidden layer consisted of three nodes and the output layer consisted of only one output node. The results from the testing data set are shown in Table 6.1.

Table 6.1: Comparison results of using Basin Average Rainfall and Five Individual Sub-basin stations.

Variables	Basin Average Rainfall	Five Individual Sub-basin Stations
EPOCHS:	7,171	8,585
CE :	99.337	99.475
RMSE :	0.263	0.220
MAE :	0.142	0.087
Raw Min:	0.208	0.208
Raw Max:	44.506	44.506
Cal Min:	0.040	-0.024
Cal Max:	43.845	43.779
Error Min:	-5.429	-4.249
Error Max:	6.762	5.028

Remarks: CE : the coefficient of efficiency; RMSE : Root mean square of error; MAE : Mean absolute error; Raw Min : Minimum measured flow; Raw Max : Maximum measured flow; Cal Min : Minimum calculated flow; Cal Max : Maximum calculated flow; Err Min : Minimum calculated error; Err Max : Maximum calculated error.

In the case with five representative sub-basin rainfalls, the rainfall inputs were increased to fifteen antecedent rainfalls, which comprised five sets of three antecedent rainfalls (RF_{t-6}, RF_{t-7}, RF_{t-8}) at each of the representative stations. The rest were kept the same as the basin average rainfall. Results for the five representative sub-basin rainfalls are also shown in Table 6.1. The results of both cases show that there is little improvement in coefficients of efficiency (CE), gained from using the distributed rainfalls over the basin average rainfall.

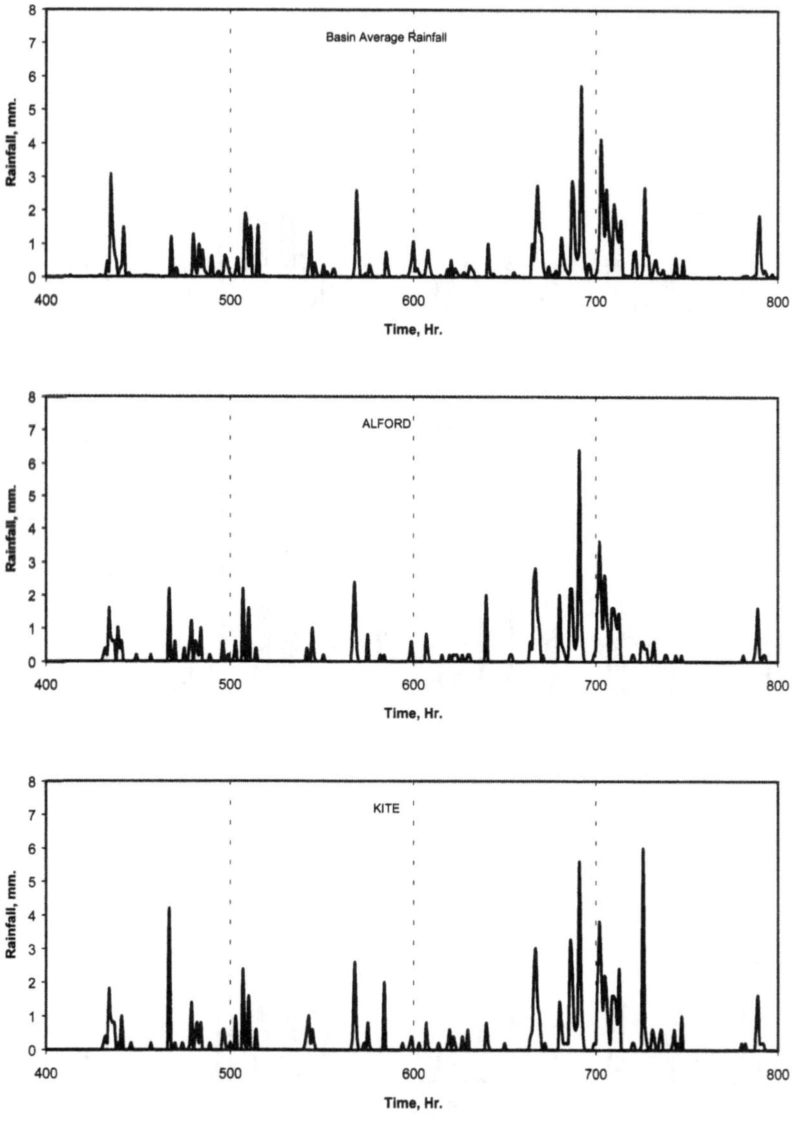

Figure 6-1 Basin Average Rainfall derived from 49 raingauges, Rainfall at stations Alford Fields and Kite Lane during the period of October 17, 1994 to November 1, 1994.

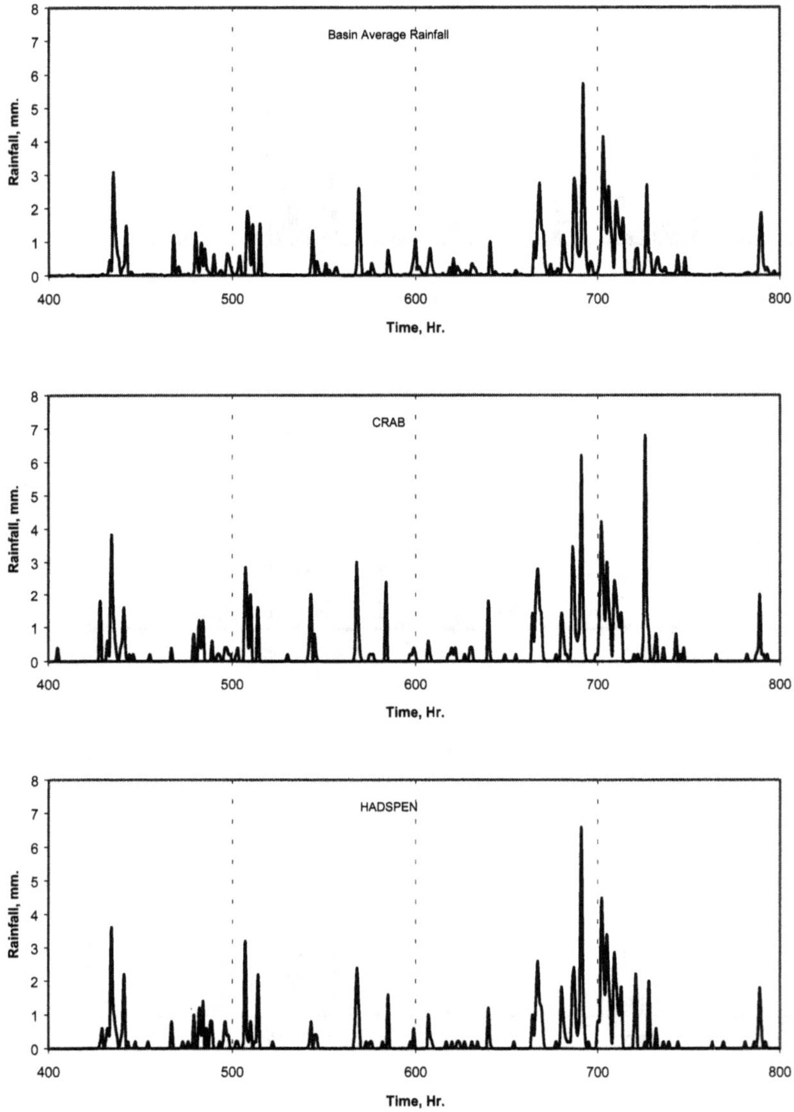

Figure 6-2 Basin Average Rainfall derived from 49 raingauges, Rainfall at stations Crabtree Land Field and Hadspen House Fields during the period of October 17, 1994 to November 1, 1994.

However, by using the distributed rainfall, the MAE (mean absolute error) has been reduced by about 50%. On the other hand, using the distributed rainfalls will increase the number of epochs by about 20%. It is obvious that a choice must be made between increased accuracy as opposed to extra computational effort. Due to the relatively small increase in accuracy in this case, it has been decided to use areal average rainfall in the current study unless otherwise stated.

6.3 Forward shift operator

In hydrological flood forecasting, the concept of the unit hydrograph (UH), initially proposed by Sherman (1932), and principally reformulated by Nash (1957) and Dooge (1959, 1973) is one of the most widely used approaches. Sempere Torres et al (1992) and Duband et al (1993) studied the river Gardon d' Anduze basin of 545 km^2, in the Cevennes, South of France using a modern version of the unit hydrograph approach. This river is subjected to flash floods. Those authors applied the unit hydrograph instead of distributed physically-based rainfall-runoff models since the UH requires less operational time. The authors derived the effective rainfall (RE) and the unit hydrograph (UH) by using the concept of *first difference transfer functions* (FDTF). The First Difference Transfer Function (FDTF) method is one of several approaches whose purpose is to identify simultaneously the linear transfer function (i.e. the finite-period unit hydrograph) and the hyetograph of effective rainfall from a sequence of storm events. A comprehensive description of the method has been provided by Duband et al (1993). The identification procedure leads only to a sequence of effective rainfall ordinates, and in order to develop an operational flood forecasting scheme a further step is required in which the series of total and effective rainfall ordinates are employed to identify a loss model. This process has been illustrated by Sempere Torres et al (1992), who evaluated three different types of loss function of differing complexity, unfortunately with only limited success.

A particular feature of the FDTF approach is the use of first differences for both the flow outputs and the transfer function. The principal advantage of first differences is claimed to be the avoidance of baseflow separation since the differencing process serves to filter out much of the slower variation in flows. Moreover, discharge increments tend to display less serial correlation than the discharges themselves, although first differencing does tend to amplify random noise in the data. For the present purpose, the FDTF method can provide a first estimate of the discharge increments over the forecast horizon and thereby supply information over and above that which an ANN can extract from data recorded up to and including the time at which the forecast is made. These discharge increments are in fact the *forward shift operator*, which provide useful information to the network and can easily be derived beforehand. The magnitude and sign of the discharge increments were anticipated to be particularly helpful in avoiding the lag of one time interval that is inherent in ANN forecasts based on recorded data only, especially at the time of occurrence of the peak discharge. In effect, the FDTF provides a first estimate of the future flow, which the ANN assimilates and corrects to provide an improved estimate.

In any operational flood forecasting scheme based on ANNs that employed the FDTF approach to provide forecasts of future flow increments, the need to identify a suitable loss model would remain. However, since the FDTF flow increments are not definitive, a comprehensive model of areal losses over the catchment is not as imperative, and a simple model would suffice. Since the purpose of the following investigation is to illustrate the principle, no attempt has been made to develop loss models for the catchments whose data are analysed. The effective rainfall sequences convolved with the identified transfer function that are employed in the ANN experiments are therefore those obtained during the original application of the FDTF algorithm.

6.4 Implementation of the FDTF Method

The FDTF method is intended to be applied to a set of rainfall events comprising the areal average rainfall totals over successive time intervals and the concurrent flows at the catchment outlet. Duband et al (1993) recommend that a minimum of 10-12 events should be selected, but that 40-50 are more appropriate for operational purposes. Given the total rainfalls and flows, there are two sets of unknowns to be identified: the average transfer function that applies to all events, and the sequences of effective rainfalls for the individual events. In effect, two sets of information are being extracted from one set of rainfall and flow series. Non-uniqueness is avoided by applying plausible constraints on the identified unknowns, i.e. volume conservation and positivity of excess rainfall totals. The solution of the problem requires an iterative alternating scheme in which one of the two sets is first assumed to be known and a solution is found for the other. For example, the excess rainfall ordinates are frozen and a solution is found for the average transfer function. After smoothing and applying the necessary constraints to the transfer function, i.e. unit volume under the function, a backward computation is performed to identify the new sequence of effective rainfall totals. The procedure continues until a criterion of convergence is satisfied. As noted previously, the computations are applied to the first differences of the flows, primarily to avoid the need for baseflow separation.

The number of catchment average rainfall totals for each event defines the maximum possible number of the (non-zero) effective rainfall totals. The number of discharge time steps is equal to the duration of the rainfall sequence plus the base length of the transfer function minus one. Indeed, the number of time intervals defining the transfer function has to be chosen prior to the computation, and is generally based upon the length of the recession (see Figure 6-3).

The starting time of an event is based on the starting time of the rising limb of the flood hydrograph. Since the convolution of the rainfall excess and the transfer function is based upon the increments in discharge, the following relationship applies:

$$q_i^L \approx \sum_{j=1}^{K+1} h_j RE_{i-j+1}^L \qquad (6.1)$$

where $q^L{}_i$ is the ith ordinate of the discharge increments in event L, h_j is the jth ordinate of the first difference transfer function, $RE^L{}_j$ is the jth ordinate of the effective rainfall for event L, and K is the number of ordinates in the FDTF.

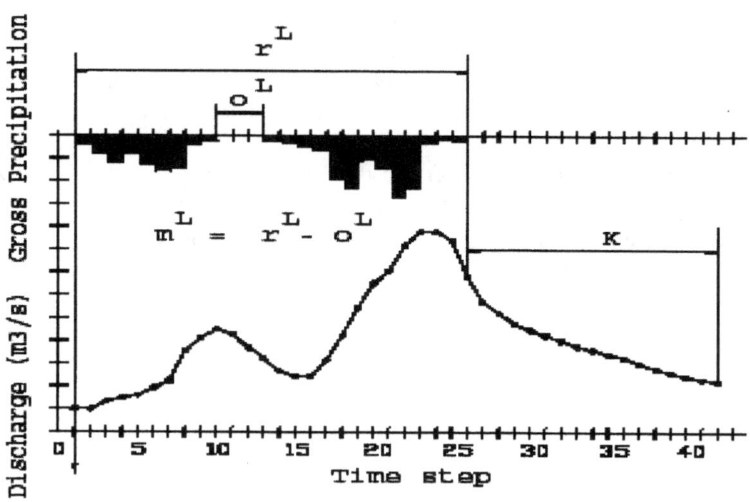

Figure 6-3 Values to be determined for an event: K = transfer function length; r^L = rainfall sequence for the event L; m^L = number of excess rainfall to be identified; o^L = number of steps without rainfall in the rainfall sequence. (Adapted from Duband et al., 1993)

The iterative identification scheme outlined above is started by assuming the total rainfall ordinates as the first approximation of the effective rainfall totals. Given *N* storm events, and using Equation (6.1) to estimate each discharge increment, results in an over-determined set of linear equations with only *K* unknowns that can be solved by the method of least squares:

$$q = RE.h \tag{6.2}$$

$$h = \left[RE^T.RE\right]^{-1} RE^T.q \tag{6.3}$$

The transfer function so identified is subject to the constraint of unit volume. The second iteration now involves the estimation of the *RE*, which can be carried out on an event basis:

$$q = F.RE \tag{6.4}$$

$$RE = \left[F^T.F\right]^{-1} F^T.q \tag{6.5}$$

where *F* is a matrix whose columns contain the FDTF ordinates, each shifted by one time unit. For each event, an over-determined set of linear equations has again to be solved, but application of a classical least squares approach can lead to instability because of the structure of the *F* matrix. Duband et al (1993) therefore proposed that in order to improve the conditioning of the system the technique of ridge regression should be applied. Firstly, the diagonal term is increased by a ridge factor of 10% in the first iteration and then reduced to zero in the later iteration. The resultant effective rainfall totals are then subject to the constraint of positivity.

The iteration continues with the newly-identified effective rainfall ordinates being used to obtain a second estimate of the transfer function. In practice, convergence is achieved in 3 to 5 iterations (Duband et al, 1993), thereby providing sets of RE^{L}_{j} and h_j that are capable of both reproducing the observed discharges and maintaining the properties of the transfer function.

To derive the effective rainfalls (REs) and the unit hydrographs (UH) for both the Brue River and the Mole River (see Appendix I for more details), the FDTF was utilised. The difference unit hydrograph (DUH) is computed by subtracting the one-time-unit forward shifted UH from the raw UH. Being insensitive to the baseflow separation, this DUH is then convolved with the effective rainfall to form the difference flows (DQ).

6.4.1 Case I : Brue river

To derive the effective rainfalls (REs) and the unit hydrographs (UH), the rainfall and discharge data for the periods of January 1, 1994-July 31, 1994 and October 1, 1994-July 31, 1995 were analysed using the FDTF method. The plot of those total rainfall (RF) and effective rainfall (RE) is shown in Figure 6-4, which demonstrates the consistency of the method to calculate the REs. The example of time series plot of 150 ordinates of the total rainfall (RF), the effective rainfall (RE) and discharge (Q mm.) during January 1994 is given in Figure 6-5. Part of the rain had wetted the basin before giving the first peak that resulted from the long duration rainfall. Then the second lower peak occurred due to the second storm that was short duration with high intensity.

The 30-minute unit hydrograph obtained for the Brue River is shown in Figure 6-6. It was found to have a time-to-peak of 9 hours. Therefore, the leading time for forecasting is bounded to be less than or equal to this value. The Jordan recurrent network (JRN) was selected to make the forecasts. All data and the difference flows were at 30 minutes time interval as suggested by T_s in Table 3.1.

The architectures of the JRNs for predicting one hour ahead (QP_{t+2}; note-two 30-min intervals) were 5-3-1. There were two inputs in the input layer consisting of the difference flows (DQ_{t+2}) and the concurrent flow (Q_t) plus the context unit, which was composed of three recurred outputs (QP_{t+1}, QP_t, QP_{t-1}). In all cases, the hidden layer consisted of three nodes and the output layer consisted of only one output node. The simple diagram of the layout of the Jordan recurrent network is given in Figure 6-7.

To march forward in time, after the first hour forecasting, the concurrent flow (Q_t) would not provide enough information for the network. Therefore, the forecasted flows were used instead. In this case, the inputs for the two hours ahead forecasting (QP_{t+4}) were the difference flows (DQ_{t+4}) and the previous forecasted flows (QP_{t+2}) plus the context unit, which was composed of three recurred outputs (QP_{t+3}, QP_{t+2}, QP_{t+1}). The rest were formulated in the same manner.

Table 6.2 shows all the forecasted results up to 12 hours ahead. It is interesting that the CEs are very well maintained above 90%, when the lead time is less than the time-to-peak of the UH, which is 9 hours. The CE degrades gradually in a non-linear way and is higher than or about 95% when forecasting is made up to 6 hours ahead.

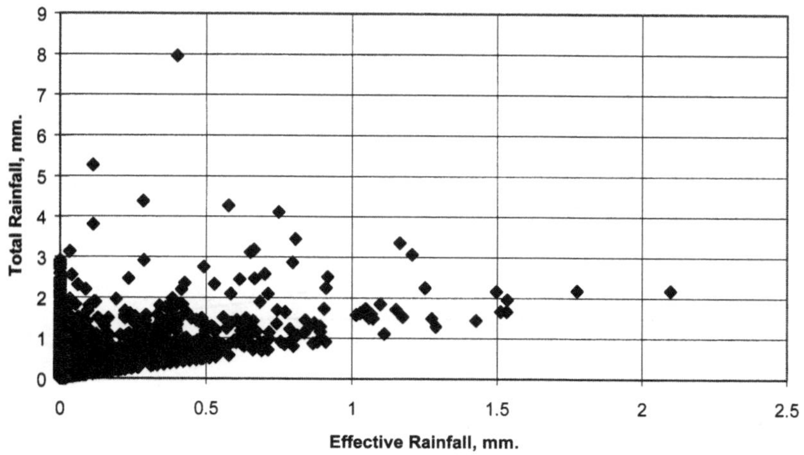

Figure 6-4 Scattergram of the total and effective rainfall obtained from FDTF method.

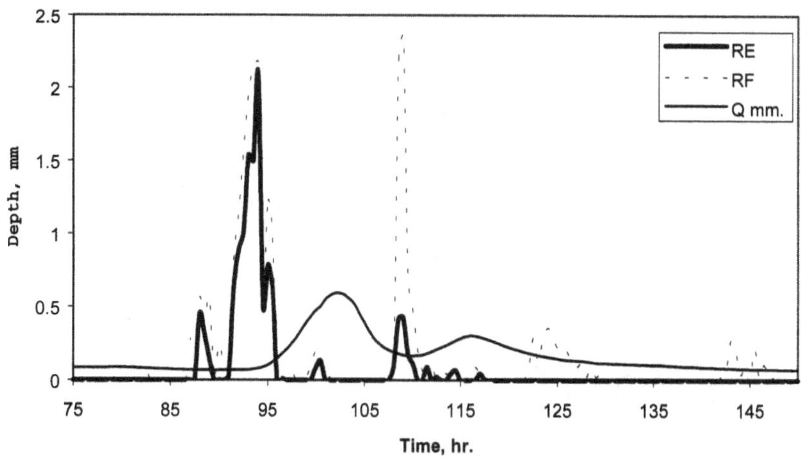

Figure 6-5 Time series plot of the total rainfall, the effective rainfall and the discharge of river Brue for the period of January 1994; RE : effective rainfall in millimetre; RF : total rainfall in millimetre;Q mm: flow in millimetre.

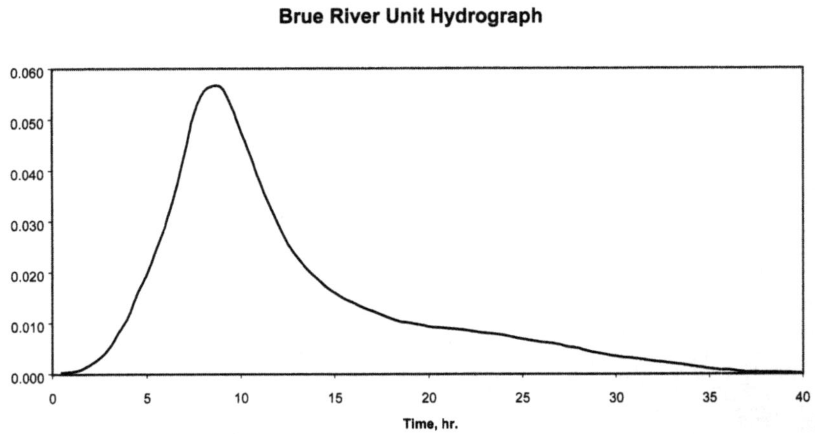

Figure 6-6 30-minute Unit Hydrograph of River Brue obtained from FDTF.

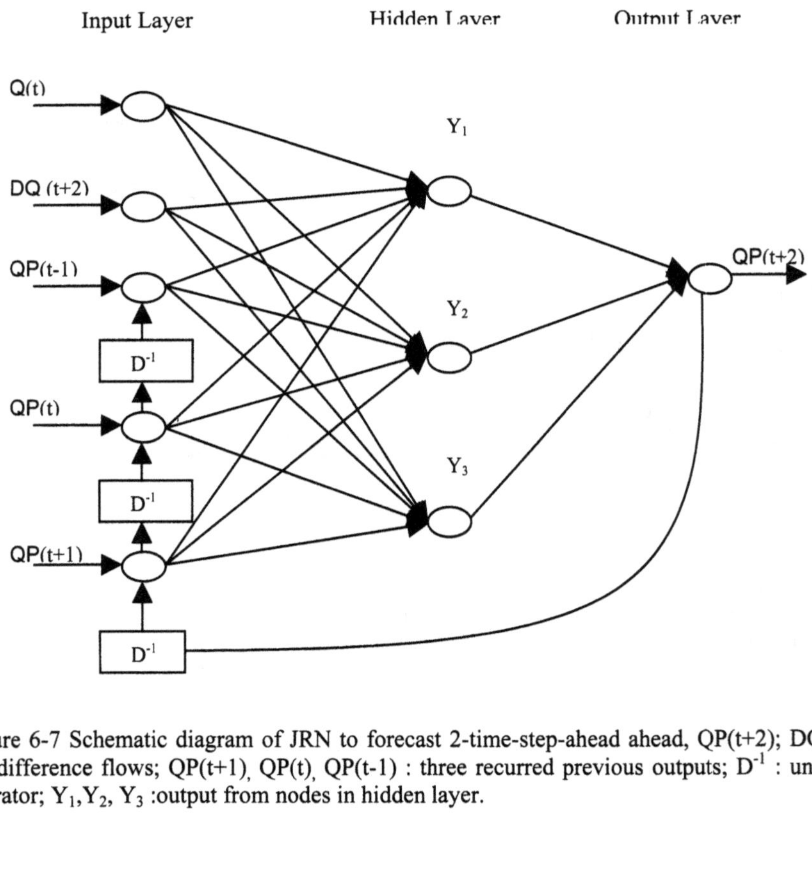

Figure 6-7 Schematic diagram of JRN to forecast 2-time-step-ahead ahead, QP(t+2); DQ(t+2) : the difference flows; QP(t+1), QP(t), QP(t-1) : three recurred previous outputs; D^{-1} : unit delay operator; Y_1,Y_2, Y_3 :output from nodes in hidden layer.

The reasons for this are due to the imperfections of the difference flows (DQs) used in calculations, as compared to the recurred errors discussed in section 5.5, and the accumulated errors inherent in the previous forecasted outputs used in the context unit.

The advantages of using the difference flows are (1) to suppress the phase errors and not allowing them to expand in time, and (2) to provide extra information to the networks. To show some examples of the results, the one and the six hour ahead forecasted results were selected and plotted. Figure 6-8 and 6-9 show the segment of 90 ordinates and the scattergram of one hour ahead forecasted flows of the Brue river using 30 minutes data interval, respectively. Figure 6-10 and 6-11 show the segment of 90 ordinates and the scattergram of six hours ahead forecasted flows of the Brue river using 30 minutes data interval, respectively.

As can be seen in Figure 6-8 and 6-10, the phase errors have been suppressed for both one and six hour ahead forecasting, and the peak flows are very well captured. It is the imperfections of the difference flows and the accumulated errors inherent in the previous forecasted outputs used as recurred inputs that cause the fluctuating behaviour in the peak flow hydrograph and in the recession.

Figure 6-9 and 6-11 show the scattergram of one and six hours forecasted and measured flows of river Brue, (QP1, QP6 and Qmeas, respectively). In general, the forecasted flows are close to the line of equal values, but tend to be more dispersed about the line when forecasting further in time. The longer the forecasting time horizon, the more the dispersion of the estimated flows.

Table 6.2: Results of flood forecasting of Brue river using JRN with the forward shift operators.

Forecasting Time (hrs.)	CE	RMSE	MAE
1	99.86	0.237	0.218
2	99.58	0.363	0.325
3	98.88	0.426	0.333
4	97.77	0.567	0.438
5	97.00	0.690	0.564
6	95.03	0.698	0.497
7	93.91	0.802	0.616
8	92.03	0.845	0.607
9	91.92	0.906	0.695
10	89.55	1.128	0.788
11	89.44	1.012	0.668
12	86.76	1.113	0.789

Remarks:CE : the coefficient of efficiency; RMSE : Root mean square of error; MAE : Mean absolute error.

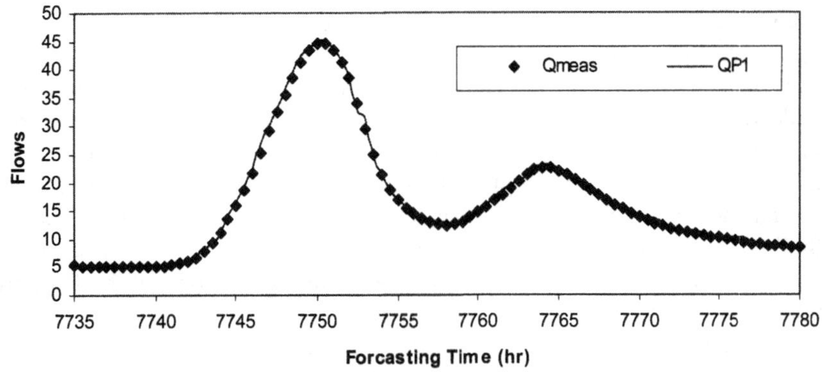

Figure 6-8 Segment of 90 ordinates of the forecasted results of one hours ahead of the Brue river using 30 minutes data interval; Qmeas : measured flows; QP1 : forecasted flows.

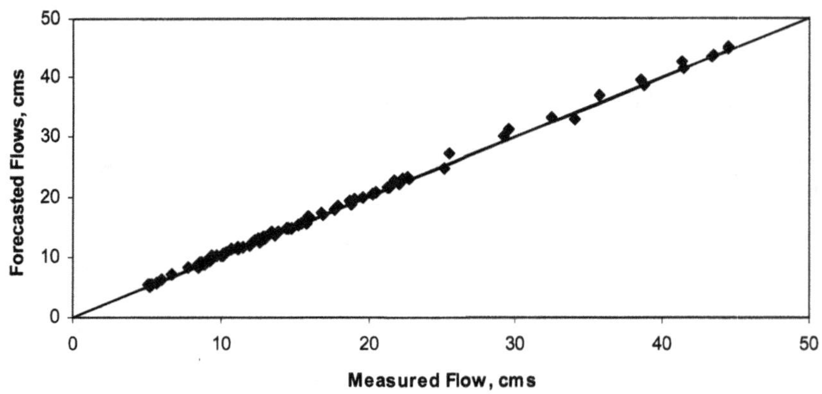

Figure 6-9 Scattergram of the one hour forecasted flows and measured flows of the Brue river using 30 minutes data interval.

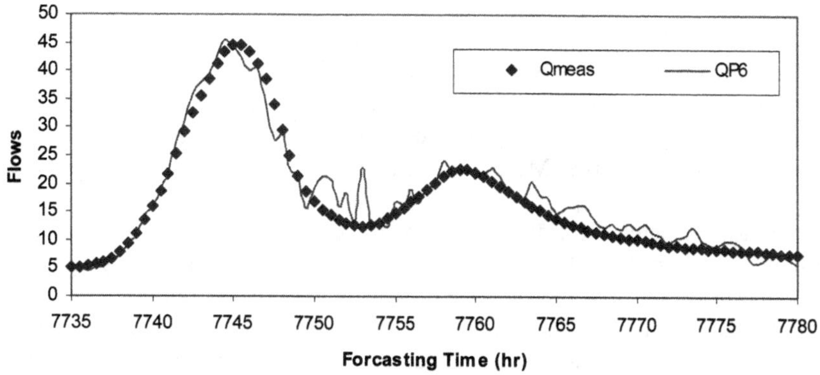

Figure 6-10 Segment of 90 ordinates of the forecasted results of six hours ahead of the Brue river using 30 minutes data interval; Qmeas : measured flows; QP6 : forecasted flows.

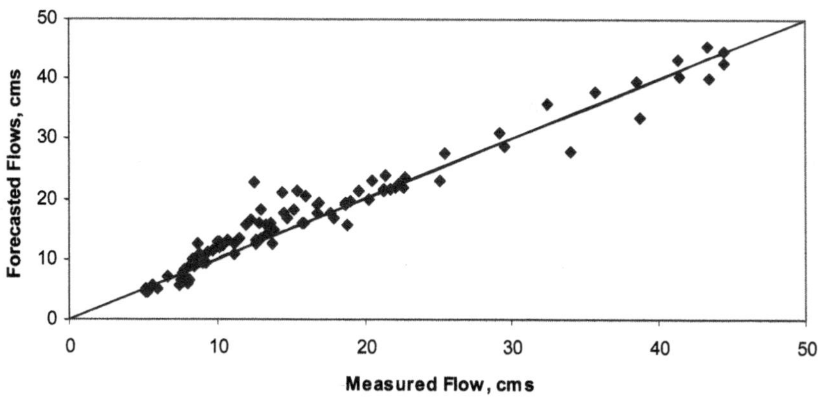

Figure 6-11 Scattergram of the six hour forecasted flows and measured flows of the Brue river using 30 minutes data interval.

6.4.2 Case II : Mole river

Dawson & Wilby (1998) studied the Rivers Amber and Mole in the UK by using feed forward neural networks. All data of both rainfalls and flows are available at 15 minutes time intervals, but only those records of the River Mole are complete for the whole year. Therefore the river Mole is selected to verify the concept of the *forward shift operator*.

Those authors selected the last 100 days of the data (autumn and early winter) for training the network and the first 100 days of the data (winter and early spring) for testing. In this study, the same data sets were used for comparison purpose.

The Mole rainfall data are measured only at the Burstow raingauge, unlike the River Brue where there are 49 raingauges distributed uniformly within a catchment about the same size as the Mole. These rainfall data may not be representative of the Mole catchment and furthermore, there are missing records in the last 25 of the first 100 days of the whole period which are used for the testing data set. The effective rainfalls (REs) and the unit hydrograph (UH) were derived by utilising the FDTF. The 30-minute UH obtained, shown in Figure 6-12, has a time to peak of 11.5 hours. In contrast to the Brue, this UH contains double peaks which are probably due to the unrepresentative basin rainfall. The same architecture of JRN used for the Brue River is adopted to make the forecasts. All data and the difference flows were also 30 minutes interval as suggested by T_s in Table 3.1.

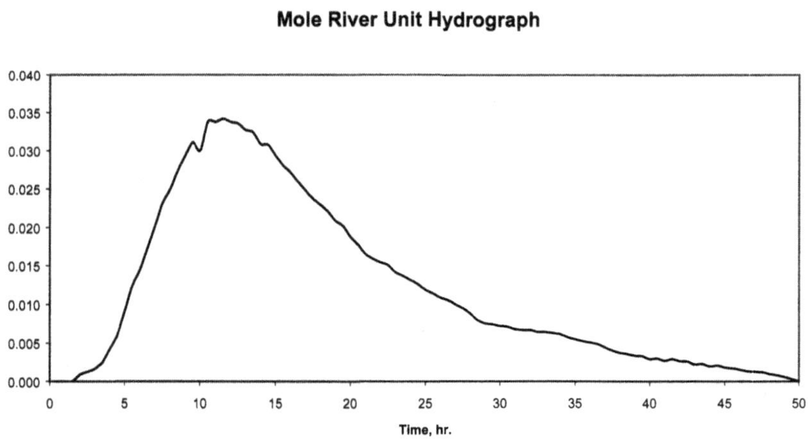

Figure 6-12 30-minute Unit Hydrograph of the Mole river obtained from FDTF.

Table 6.3 shows all the forecast results up to 12 hours ahead. In this case the CEs are very well maintained above 90% when the lead-time is less than 7 hours. The CE degrades rapidly in a non-linear way in contrast to the River Brue. The reasons for this behaviour are due mainly to the rainfall data, which were obtained from a single gauge, the imperfections of the difference flows (DQ) derived from the doubtful rainfall data, and the accumulated errors inherent in the forecasted outputs used in the context unit.

Table 6.3: Results of flood forecasting of Mole river using JRN with the forward shift operators.

Forecasting Time (hrs.)	CE	RMSE	MAE	MSRE
1	99.87	0.243	0.120	0.002
2	99.00	0.657	0.422	0.023
3	97.71	1.029	0.627	0.069
4	97.08	1.326	0.953	0.154
5	94.17	1.527	0.638	0.039
6	93.54	1.603	0.787	0.093
7	90.70	1.944	0.867	0.077
8	89.25	2.046	0.903	0.093
9	87.47	2.265	1.180	0.196
10	86.76	2.284	1.164	0.196
11	84.12	2.608	1.497	0.343
12	81.52	2.778	1.526	0.341

Remarks: CE : the coefficient of efficiency; RMSE : Root mean square of error; MAE : Mean absolute error; MSRE : Mean square relative error.

The one and the six hour ahead forecasted results are shown in Figure 6-13 & 6-14 and Figure 6-15 & 6-16, respectively. Figure 6-13 and 6-15 show a segment of 750 ordinates of one- and six-hours-ahead forecasted flows of the Mole river using 30 minute data, respectively. Figure 6-14 and 6-16 show the associated scattergram.

To show the effects of the missing rainfall records in the last 25 days of the testing data set, Figure 6-17 & 6-18 show the segment of the last 600 ordinates of one and six hours ahead forecasted flow. The phase errors are pronounced in the case of the six-hour forecasting because of the lack of extra information provided by the difference flows. Therefore, the quality of the forecasts is poorer when the lead time is advanced.

As shown in Figure 6-13, 6-15, 6-17 and 6-18, the phase errors have partly been suppressed for both one- and six-hour ahead forecasting. The peak flows are well captured where there are rainfall data available. The imperfections of the difference flows and the accumulated errors in forecasted outputs used in the context unit cause the fluctuations in the peak flow hydrograph.

Figure 6-14 and 6-16 show the scattergrams of one- and six-hours forecasts and measured flows for the river Mole. In general, the forecasted flows lie close to the line of equal values for the first hour, but tend to disperse more along the line when the forecast time increases. The results of the Mole river are poorer than those of the Brue river.

For the six-hour-ahead forecasts, Dawson & Wilby (1998) obtained an RMSE and an MSRE of 1.906 and 0.339, respectively, compared with 1.603 and 0.093 in Table 6.3.

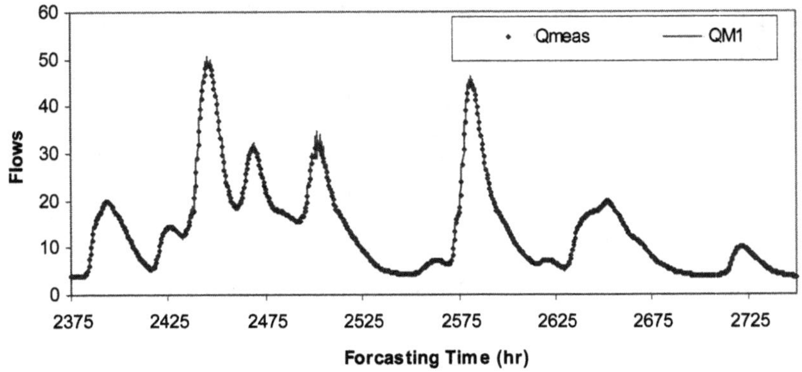

Figure 6-13 Segment of 750 ordinates of the forecasted results of one hours ahead of the Mole river using 30 minutes data interval; Qmeas : measured flows; QM1 : forecasted flows.

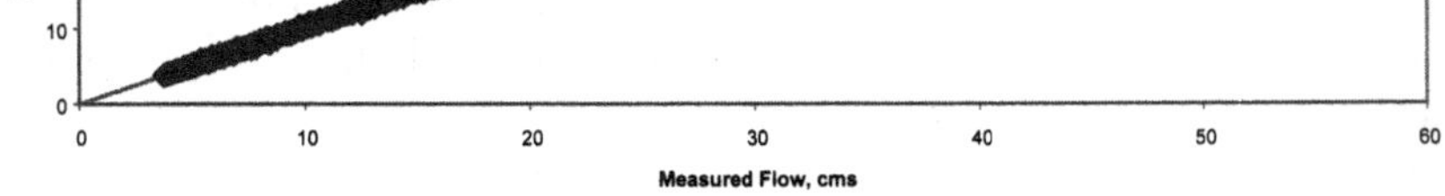

Figure 6-14 Scattergram of the one hour forecasted flows and measured flows of the Mole river using 30 minutes data interval.

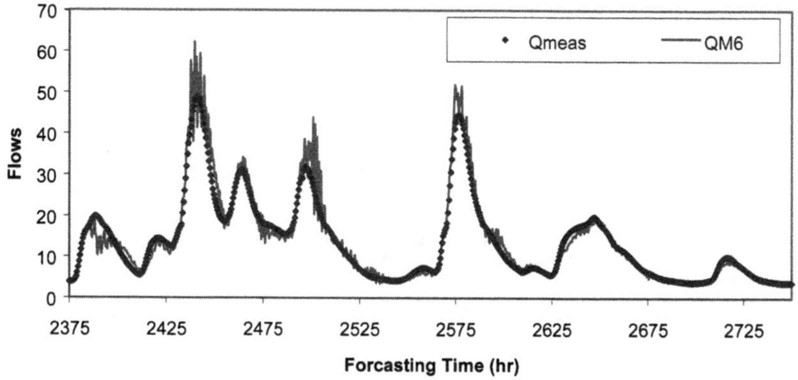

Figure 6-15 Segment of 750 ordinates of the forecasted results of six hours ahead of the Mole river using 30 minutes data interval; Qmeas : measured flows; QM6 : forecasted flows.

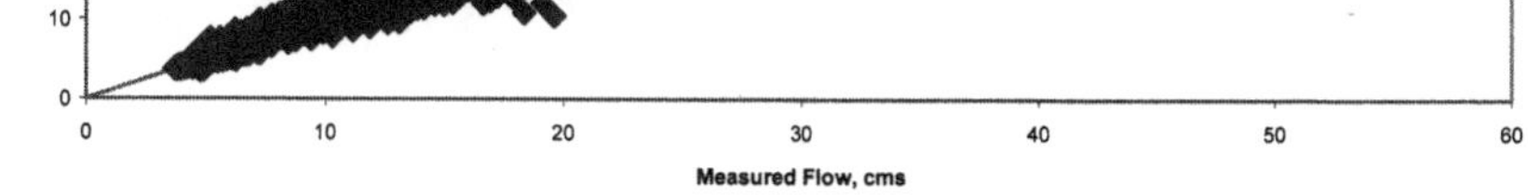

Figure 6-16 Scattergram of the six hour forecasted flows and measured flows of the Mole river using 30 minutes data interval.

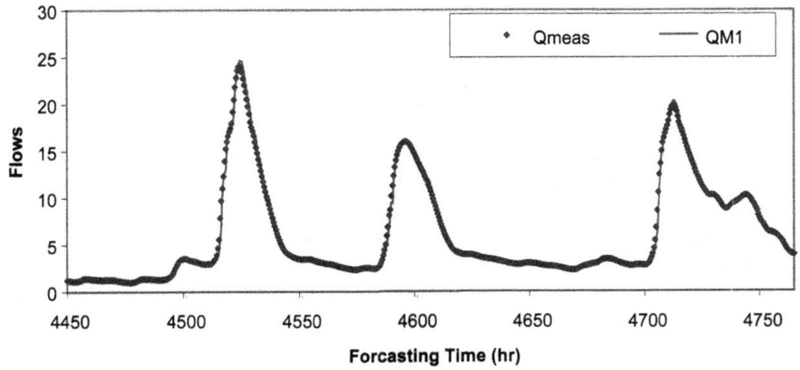

Figure 6-17 Segment of the last 600 ordinates of the forecasted results of one hours ahead of the Mole river using 30 minutes data interval, where there are no rainfall data available; Qmeas : measured flows; QM1 : forecasted flows.

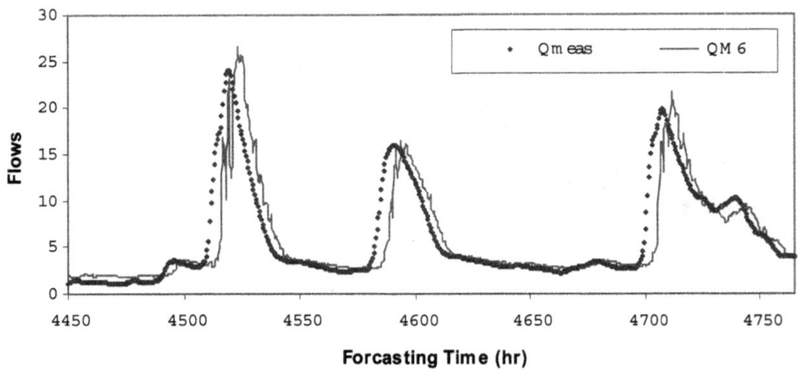

Figure 6-18 Segment of the last 600 ordinates of the forecasted results of six hours ahead of the Mole river using 30 minutes data interval, where there are no rainfall data available; Qmeas : measured flows; QM6 : forecasted flows.

6.5 Discussions and Conclusions

To obtain a good flood forecasting result, the appropriate sampling time interval T_s, should be selected carefully. The formula suggested by Sugawara (1995) provides a reasonable initial basis for this task. This formula may not be applicable everywhere in the world, therefore the refinement of this formula must be continued. The formula should try to incorporate the drainage area, the slope of the basin, the length of the main channel and various aspect of basin characteristics to make it more universal.

The use of rainfall data is one of the major aspects, i.e. should the rainfall input be distributed or areal average rainfall. The details shown in section 6.2 reveals that the areal average rainfall not only reduces load of input data but also increases the computational speed without loss of accuracy.

Then, in order to forecast forward in time, the FDTF method is applied to provide a forward shift operator to the network. As presented in section 6.4.1 and 6.4.2 that the difference flows (DQs) derived from the difference unit hydrograph (DUH) can provide extra information to the network to produce very accurate results. In the absence of rainfall data, the phase error will be pronounced and the results will deteriorate. The spikes in the peak hydrograph and in the recession are due to the imperfections of the difference flows as compare to the recurred errors discussed in section 5.5 and the accumulated errors inherent in the previous recurred output used in the context unit.

It is much easier to obtain very accurate results by using the flow data upstream of the forecasting location, as shown in case of the Chumporn river in Chapter 5. The upstream flow data are in fact the convolution of the effective rainfall over the upstream sub-catchments, and there is no need to derive the DQ.

For smaller catchments in general, there are only rainfall and flow data available without any upstream flow data. For this reason, the FDTF is proposed to derive both the effective rainfall (RE) and the unit hydrograph (UH) which finally will lead to the difference flow (DQ). It is the DQ that provides extra information to the network.

7

Conclusions and recommendations

The ability to provide sufficient advance warning of flood occurrence is fundamental to reducing the potentially disastrous effects of floods (Ford, 2001). Sufficient advance warning time may save lives by giving floodplain residents time to remove themselves and their possessions to safety, and may save property by allowing time to effect various structural and other adjustments. Earlier advanced warning can be achieved through advances in mathematical modelling, the installation of telemetry and field monitoring equipment at critical sites in drainage networks, through satellite and radar sensing of extreme rainfalls, and through the coupling of precipitation and runoff models (Takeuchi, 2001; Jasper et al, 2002). However, in practice, successful real-time flood forecasting often depends on the efficient integration of all these separate activities.

There are many different types of models for the purposes of flood forecasting, which range from purely deterministic models to hybrid stochastic-deterministic models. They vary according to their data requirements, the computational resources that are necessary for their implementation, and the problem to be addressed. In general, calibration of these models is not an easy task. Some prior knowledge is required of physical processes taking place in the catchments. The conceptual models generally have a large number of parameters which are not directly measurable and must be estimated through model calibration and verification.

The overview of forecasting methods in Chapter 1 serves to illustrate the wide variety of methods currently applied to the problem of flood forecasting. In recent years, the introduction of weather radar and quantitative precipitation forecasting has begun to create new opportunities for the development of more sophisticated flood forecasting systems. However, with the possible exception of a few notable cases of large trans-boundary rivers, these advanced systems are most unlikely to be deployed in developing countries within the near future. There remains a clear need for forecasting methods that require a minimum of data for calibration and operation, can perform in real time and yet can deliver reliable and accurate forecasts for a suitable range of forecast horizons. Data driven methods have been identified as possessing the potential to fulfil these needs, and therefore provided the primary focus of this work. In particular, the recent introduction of Artificial Neural Networks (ANNs) as hydrological models has offered the opportunity to relate directly the rainfall and the flow without requiring some empirical prior data processing in the form of baseflow separation and determination of the proportion of rainfall that is effective in producing runoff. ANNs offer the opportunity to avoid such prior processing since they operate without the constraint of volume continuity. Furthermore, the complex nature of hydrological systems can be modelled through non-linear relationship of ANNs.

However, ANNs have found to have inability to extrapolate (Hall & Minns, 1993). The input to each node in the hidden layer of a multi-layer perceptron type of ANN consists of a summation of the products of the weights assigned to each connection from the input layer and the inputs themselves (see Figure 2-2). This sum is then passed through an activation function, generally of sigmoidal form, such that the outputs from the hidden layer are constrained to a range from zero to one. The nodes in the output layer perform a similar operation. The result is that the extremes in their training data set are scaled to the zero-one interval. If then (say) the largest values in the testing data set exceed those in the training set, then the former will not be reproduced by the model. Some latitude is provided by varying the range of standardisation to, for example, 0.2-to-0.8, but the ability of the ANN to extrapolate remains limited (Hettiarachchi et al, 2003). Although the use of a linear activation function in the output layer may appear to offer a degree of extrapolation beyond fixed bounds, in practice the amount of extrapolation is limited by the saturation of the nodes in the hidden layer (Karunanithi et al, 1994).

In this present work, three main points in applying ANNs in flood forecasting schemes were addressed.

- Firstly, the extrapolation problem inherent in a network.
- Secondly, the advantage of different network architectures such as the feed forward network and the recurrent network. Two main recurrent networks, namely the Elman and Jordan schemes were studied in depth.
- Lastly, the choices of inputs that will give the best results. The overall objective was to provide accurate forecast results and lengthen the lead-time.

A general introduction to flood forecasting and then the history and a review of the state of the art of applying Artificial Neural Networks are giving in Chapter 1. The three problems involved in the use of ANN were also outlined.

In Chapter 2, the basic concept of ANNs e.g. the biological neuron and the artificial neuron or perceptron is given. The feed-forward network was introduced along with its learning rule and various activation functions. Detailed analysis of these activation functions in terms of their functional values and derivatives were given. These analyses form the basis for solving the extrapolation problems faced in the past. Then the recurrent network algorithms, the Elman and Jordan schemes were described at the end of the Chapter.

Chapter 3 describes the preliminary considerations which are necessary for the study. These concern the choice of a suitable time interval and the selection of a goodness-of-fit criterion for modelling. The choice of a suitable time interval has simply been ignored by many researchers in the past. Too long a time interval will smooth out the essence of the response to storm rainfall. Therefore, accurate flood forecasts cannot be issued. Childs (1982; see Chapman, 1994) and Sugawara (1995) have suggested empirical formulae for a suitable time interval for analyses. To see the effects of different time interval based on Sugawara's criterion, the data of river Brue were resampled into different intervals started from 30 minutes to 1, 2, 3, 4, 6, 12 and 24 hours. Then these different data were input to a feed forward neural network to predict one time step ahead. The results show that the more the value of time interval deviates from the criterion, the worse the forecast results. Therefore, the formula given by Sugawara (1995) was employed throughout this study. The other problem of goodness-of-fit was addressed in the same chapter. Various indices attempt to describe different errors in the model, e.g. the volumetric error, the peak error and the time-to-peak error. It was pointed out that there is no universal index to work with in order to describe the goodness-of-fit of the model (Diskin & Simon, 1977). Therefore to apply such an index depends upon the purpose of producing the results. In this study, the coefficient of efficiency introduced by Nash and Sutcliffe (1970) was employed throughout, largely because of its familiarity. However, where appropriate, this index was supplemented by other criteria.

Although forecasters have taken advantage of conceptual rainfall-runoff models in many previous applications, the use of data-driven methods has only been the subject of serious investigation within the last five to ten years. Of these methods, ANNs have received the most attention, yet have suffered most obviously from the problems of extrapolation. Despite the attention which has been drawn to the extrapolation problem, work continues to be published in which its solution is not addressed (see, e.g. Cameron et al, 2002). In Chapter 4, a method of creating room for an ANN to extrapolate has been proposed, which focuses on the setting of the range of the activation functions applied to each node in the network. This approach has been successfully applied to several case studies, i.e. the Silk Stream and Rivers Thrushel and Brue. However, the accuracy of the forecast deteriorates quickly when the forecast time horizon is extended with standard MLPs, largely because of the phase errors that become apparent. This *forward difference problem* is typically encountered when a change in flow rate occurs that was not anticipated by the ANN. A specific example is the occurrence of the peak of the hydrograph, where despite the lower rate of change of discharge with respect to time, the computed output will continue to rise even when the observed discharge begins to fall. The result is a built-in error of one time interval. Unfortunately, such phase errors are not well reflected in the values of coefficients of efficiency (see Hall, 2001), but are readily apparent on visual inspection.

Because the simple architecture of an MLP resulted in poor performances at forecast time horizons beyond the initial one step ahead, the possibility that more sophisticated networks with outputs feeding back into the input or the hidden layers was investigated in Chapter 5. The 'memory' of training events that is inherent in the Jordan network was found to provide some improvement in the performance over the standard MLP networks, but the 'forward difference problem' was still evident.

An alternative approach to treating the forward difference problem is that of using the ANN to 'correct' a crude forecast obtained by another method. For the purposes of illustration, a transfer function approach was chosen to explore this alternative and was investigated in Chapter 6. The knowledge gained from the transfer function helps improving the accuracy of forecast lead times up to catchment response time.

From an operational point of view, the FDTF cannot be applied to the real time gross rainfall data in order to derive the forward shift operator, since the corresponding flows are not available. Furthermore, this operator depends on the loss function used to derive the effective rainfall and the unit hydrograph, which are specific to each river basin. Therefore, the appropriate loss function must be obtained from the observed gross rainfall, with an output as close as possible to the effective rainfall identified by the FDTF. Once the appropriate loss function is known, the predicted effective rainfall can be derived and be convolved with the difference unit hydrograph to produce the predicted difference flows used as the forward shift operator to the network. From these predicted difference flows, the forecasting can then be issued.

In order to increase the flood warning lead time, rainfall forecasting models should be incorporated in the rainfall-runoff models. There is a possibility to incorporate the forecasted rainfall either from radar sensing of extreme rainfall, satellite, and/or the large scale precipitation forecast (Hu, 2002) in the ANNs model by using the FDTF. A particular problem with the use of meteorological forecast products such as those from the European Centre for Medium-range Weather Forecasting (ECMWF) employed by Hu (2002) was the clear improvement in the quality of the forecasts over time. Such inconsistencies pose particular difficulties in constructing sufficiently large and representative data sets for training, cross validation and testing of ANN models. A further disadvantage of some of those rainfall forecast products is their grid size resolution. For example, the ECMWF large scale precipitation forecast has a resolution of 60 km, which is too coarse for small and medium sizes of catchment. Further development of instrument and technique will be required.

In summary, this work with Artificial Neural Networks has demonstrated that ANNs may be used to advantage as rainfall-runoff models under the category A.2 proposed by Nemec (1986). Moreover, neural networks may also be employed as routing models under category A.1, as shown in Chapter 5. ANNs have the advantages of simplicity, ease of use and rapid run times. However, the problem of parameter identification encountered in standard rainfall-runoff models is replaced by that of determining the 'pattern' of inputs that produces a 'pattern' of outputs. In Chapter 4, a method of overcoming the 'extrapolation problem' of neural networks has been suggested. If an ANN has the room to extrapolate, then attention may be concentrated on the ability to forecast one or more time steps ahead. Chapter 5 has demonstrated that recurrent network architectures, particularly Jordan networks, can improve upon the performance of networks without feedback, such as multi-layer perceptrons. However, the additional 'memory' of Jordan networks only partially helps with overcoming the 'forward difference problem'. The use of ANNs to correct a crude forecast, demonstrated in Chapter 6, has been shown to be more successful. However, as Sempere Torres et al (1992) have already indicated, further work will be required on the selection of an appropriate loss function for a given catchment before this approach can be fully implemented.

8

Samenvatting

De mogelijkheid voldoende vooraf te waarschuwen voor overstromingen is zeer belankrijk voor het reduceren van de potentieel desastreuze effecten hiervan (Ford, 2001). Het tijdig evacueren van mensen en goederen, evenals het aanbrengen van structurele noodaanpassingen kan levens en eigendommen sparen. Eerdere waarschuwing kan worden bereikt door verbeteringen in de mathematische modelering, het installeren van telemetrie en veldwaarnemings apparatuur op kritische plaatsen in drainage netwerken, sateliet and radar waarneming van extreme neerslag, en het koppelen van neerslag en afvoer modellen (Takeuchi, 2001; Jasper et al., 2002). Succesvolle real-time overstromingsvoorspelling hangt in de praktijk echter vaak af van het efficient integreren van deze activiteiten.

Er zijn veel verschillende types van modellen om overstromingen te voorspellen, varierend van zuiver deterministisch tot hybride stochastisch-deterministisch. De onderlinge verschillen hebben betrekking op de gegevensbehoefte, de benodigde rekenfaciliteiten en de aard van het op te lossen probleem. In het algemeen is de calibratie van deze modelen geen eenvoudige taak. Een zekere mate van voorkennis van de hydrologische processen in een stroomgebied is noodzakelijk. Conceptuele modellen hebben in het algemeen een groot aantal niet direct meetbare parameters welke bepaald moeten worden door calibratie en verificatie.

Het overzicht van voorspellingsmethoden in Hoofdstuk 1 dient om de varieteit in momenteel toegepaste overstromingsmodellen te illustreren. De introductie van weerradar en quantitatieve neerslagvoorspelling heeft moglijkheden geschapen voor de verdere ontwikkeling van overstromings voorspellingssystemen. Echter, met mogelijke uitzondering van enkele grote grensoverschreidende rivieren zullen deze geavanceerde systemen waarschijnlijk niet in de nabije toekomst in ontwikkelingslanden worden toegepast. Er blijft duidelijke behoefte aan methoden welke een minimum aan gegevens vragen voor calibratie en verificatie, welke in real-time kunnen opereren en tevens betrouwbare voorspellingen kunnen leveren. De data-driven methoden kunnen mogelijk aan deze voorwaarden voldoen en vormen het primaire aandachtspunt van het huidige werk. In het bijzonder, de recente introductie van kunstmatige neurale netwerken (artificial neural networks, ANNs) als hydrologische modellen bieden de mogelijkheid om neerslag en afvoer direct aan elkaar te relateren, zonder dat voorafgaande gegevensverwerking nodig is in de vorm van basisafvoerseparatie en bepaling van effectieve neerslag nodig is. De ANNs bieden deze mogelijkheid omdat zij functioneren zonder de noodzaak van volumeconservatie. Daarenboven kan de complexe natuur van hydrologische systemen door de niet-lineaire relaties van ANNs worden gemodelleerd.

Het is echter gebleken dat ANNs niet in staat zijn om te extrapoleren (Hall en Minns, 1993). De invoer naar elk knooppunt in de hidden layer van een multi-perceptron type ANN bestaat uit een sommatie van het produkt van de gewichten toegekend aan elke verbinding van de input layer en de invoeren zelf (zie Figuur 2.2). De som wordt gebruikt in een activatiefunctie, gewoonlijk van sigmoidale vorm, zodat de uitvoer uit de hidden layer begrenst wordt tussen 0 en 1. Het resultaat is dat de extreme waarden in de training gegevensreeks worden geschaald binnen het 0-1 interval. Als vervolgens de (zeg) grootste waarden in de testgegevens die van de training gegevens overschrijden, zullen deze waarden niet gereproduceerd worden door het model. Enige flexibiliteit wordt bereikt door de standaardizering aan te passen, bijvoorbeeld van 0.2-0.8, maar het extrapolatievermogen van ANNs blijft beperkt (Hettiararchi et al., 2003). Hoewel het gebruik van een lineaire activatiefunctie in de output layer een mate van extrapolatie buiten de grenzen lijkt op te leveren, blijft deze mate beperkt door verzadiging van de knooppunten in de hidden layer (Karunanithi et al., 1994).

Drie hoofdpunten in de toepassing van ANNs ten behoeve van overstromingsvoorspelling worden in dit proefschrift aan de orde gesteld:

- Ten eerste, het extrapolatieprobleem welke inherent is in een netwerk.
- Ten tweede, het voordeel van verschillende netwerk architecturen, zoals het feed-forward network en het recurrent network. Twee belangrijke recurrent networks, te weten de Elman en Jordan schema's werden in detail onderzocht.
- Tot slot, de keuze van invoergegevens welke de beste resultaten geven. Algemeen doel hierbij was nauwkeurige voorspellingen en het verlengen van de zogenaamde voorspellingstijd.

Een algemene introductie in het voorspellen van ovestromingen, gevolgd door een overzicht van toepassingen van ANNs wordt gepresenteerd in Hoofdstuk 1. Tevens worden hier de drie problemen met betrekking tot het gebruik van ANNs behandeld.

In Hoofdstuk 2 wordt het basis-concept van ANNs, bijvoorbeeld de biologische neuron en de kunstmatige neuron of perceptron behandeld. Het feed-forward network wordt geintroduceerd, tesamen met de leer-regel en verscheidene activatie functies. Een gedetailleerde analyse van deze activatie functies in termen van functie-waarden en afgeleiden wordt gepresenteerd. De analyses vormen de basis om het extrapolatieprobleem op te lossen. Tenslotte worden de recurrent networks, de Elman en Jordan schema's behandeld.

Hoodstuk 3 beschrijft de voorafgaande overwegingen welke nodig zijn voor de studie. Deze betreffen de keuze van een geschikt tijdsinterval en de selectie van een goodness-of-fit criterium. Het geschikte tijdsinterval is in het verleden door veel onderzoekers eenvoudigweg genegeerd. Een te groot tijdsinterval zal de neerslag respons doen afvlakken. Accurate overstromingsvoorspellingen kunnen dan niet gegeven worden. Childs (1982; zie Chapman, 1994) en Sugawara (1995) suggereren empirische formulae voor de geschikte tijdsintervallen. Om het effect van veschillende Sugawara-gebaseerde tijdsintervallen zichtbaar te maken, zijn de gegevens van de Brue rivier herbepaald voor intervallen van 30 minuten, en 1, 2, 3, 4, 6, 12 en 24 uur. Vervolgens zijn deze gegevens ingevoerd in een feed-forward network om een tijdstap vooruit te voorspellen. Het resultaat laat zien dat naarmate de tijdsinterval waarde verder afwijkt van het criterium, de voorspelling slechter wordt. De formule van Sugawara (1995) is daarom voor deze gehele studie gebruikt. Het probleem van de goodness-of-fit is dat verschillende indices hiervoor veschillende soorten afwijkingen beschrijven. Bijvoorbeeld de volumetrische afwijking, de piek-afwijking en de afwijking in de tijd tot aan de piek. Het is al beschreven dat er geen algemeen toepasbare index is voor bepaling van de mate van model overeenkomst (Diskin en Simon, 1977). De te gebruiken index hangt daarom af van het doel van de toepassing. Voor deze studie is de coefficient of efficiency van Nash en Sutcliffe (1970) gebruikt vanwege de bekendheid. Niettemin zijn waar van toepassing tevens andere criteria gebruikt.

Alhoewel voorspellers de voordelen van conceptuele neerslag-afvoer modellen gebruikt hebben in veel bestaande applicaties, is de toepassing van data-driven methoden gedurende de laatste vijf tot tien jaar serieus onderzocht. Met betrekking tot deze methoden hebben ANNs de meeste aandacht gekregen, maar lijden duidelijk aan het extrapolatieprobleem. Ondanks de aandacht voor het extrapolatieprobleem, wordt nog steeds werk gepubliceerd waarin geen oplossing wordt gezocht (zie bijvoorbeeld Cameron et al., 2002). In Hoofdstuk 4 wordt een methode voor extrapolatie in ANNs voorgesteld, welke gericht is het vaststellen van een bereik voor de activatiefuncties voor elk knooppunt in het netwerk. Deze benadering is met succes toegepast op veschillende situaties, m.n. de Silk Stream, en de Thrushel and Brue rivieren. Niettemin neemt de nauwkeurigheid snel af wanneer de voorspellingstijd word vergroot in standaard MLPs, grotendeels omdat de fasefouten duidelijker worden. Dit forward difference problem treed met name op wanneer een verandering in afvoer optreed welke niet door het ANN geanticipeerd was. Een specifiek voorbeeld is het voorkomen van een piek in de afvoercurve, waarin ondanks de lagere verandering van afvoer per tijd de berekende afvoer zal blijven stijgen terwijl de waargenomen afvoer afneemt. Dit resulteert in een ingebouwde fout van een tijdstap. Helaas worden deze fasefouten niet goed weergegeven in de waarden van coefficients of efficiency, maar zijn direct duidelijk uit visuele inspectie van de resultaten.

Omdat de eenvoudige MLP architectuur slechte resultaten opleverde bij voorspellingstijden groter dan een tijdstap, werd in Hoofstuk 5 de mogelijkheden ondezocht van meer verfijnde netwerken waarin de uitvoeren worden teruggeleid naar de input of de hidden layers. Het 'geheugen' inherent aan het Jordan network bleek enige verbetering in prestatie ten opzichte van het standaard MLP netwerk te bieden, maar het 'forward difference problem' was nog steeds aanwezig.

Een alternatieve benadering om het forward difference probleem aan te pakken is het ANN een ruwe voorspelling door een andere methode te laten 'corrigeren'. Bij wijze van illustratie is een transferfunctie aanpak als dergelijk alternatief onderzocht in Hoofdstuk 6. The kennis verkregen uit de transferfunctie helpt om de nauwkeurigheid van voorspellingstijd tot aan de stroomgebieds responstijd.

Vanuit operationeel oogpunt kan de FDTF niet toegepast worden op de real time gross neerslaggegevens om de forward shift operator af te leiden, omdat de corresponderende afvoeren niet beschikbaar zijn. Bovendien hangt deze operator af van de verliesfunctie welke gebruikt wordt om de effectieve neerslag en de unit hydrograph te bepalen, die specifiek zijn voor elk stroomgebied. De geschikte verliesfunctie dient daarom bepaald te worden uit de waargenomen gross neerslag, met een uitkomst welke zo dicht mogelijk de effectieve neerslag benaderd als geidentificeerd door de FDTF. Als de geschikte verliesfunctie eenmaal bekend is, kan de effectieve neerslag bepaald en geconvolueerd worden met de verschil unit hydrograph om de voorspelde verschilafvoeren te produceren die als forward shift operator gebruikt worden door het netwerk. Vanuit deze voorspelde verschilafvoeren kan vervolgens de voorspelling gedaan worden.

Om de voorspellingstijd te vergroten zouden neerslag voorspellingsmodellen moeten worden ingesloten in neerslag-afvoer modellen. Er zijn mogelijkheden om neerslag bepaald uit radarwaarneming van extreme regenval, satelliet en/of ruime-schaal neerslagvoorspellingen (Hu, 2002) in het ANN te incorporeren door gebruik te maken van de FDTF. Een typisch probleem met he gebruik van meteorologische voorspellingen, van bijvoorbeeld het European Centre for Medium-range Weather Forecasting (ECMWF) toegepast door Hu (2002) was de duidelijke verbetering van deze voorspellingen met de tijd. Dergelijke inconsistensies maken het moeilijk om voldoende lange representatieve gegevensreeksen voor training, kruis-validatie en het testen van ANN modellen te construeren. Een verder nadeel van sommige voorspellingen kan de ruimtelijke resolutie zijn. De resolutie van bijvoorbeeld de ECMWF large scale precipitation forecast is 60 km, welke te grof is voor kleine to middelgrote stroomgebieden. Verdere ontwikkeling zal dan ook noodzakelijk zijn.

Samenvattend heeft dit werk aangetoond dat ANNs met gunstig resultaat gebruikt kunnen worden als neerslag-afvoer modellen in de categorie A.2 volgens Nemec (1986). Tevens kunnen neurale netwerken toegepast worden als routing modellen in de categorie A.1, zoals Hoofdstuk 5 laat zien. ANNs bezitten de voordelen van eenvoud, gebruiksgemak, en snelle rekentijden. Daarentegen wordt het probleem van parameterbepaling in standaard neerslag-afvoer moellen vervangen door dat van het bepalen van het invoerpatroon welke het uitvoerpatroon genereerd. In Hoofdstuk 4 wordt een methode om het extrapolatieprobleem aan te pakken gesuggereerd. Indien een ANN de ruimte heeft om te extrapoleren, kan de aandacht gericht worden op de mogelijkheid om een of meerdere tijdstappen vooruit te voorspellen. Hoofstuk 5 toont aan dat recurrent network architecturen, in het bijzonder Jordan networks, in staat zijn om de prestaties van netwerken zonder terugkoppeling, zoals multi-layer perceptrons, te verbeteren. Het additionele 'geheugen' van Jordan networks helpt echter slechts gedeeltelijk om het forward difference problem te overkomen. Het gebruik van ANNs om een ruwe voorspelling te corrigeren is meer succesvol, zoals gedemonstreerd in Hoofdstuk 6. Zoals niettemin al door Sempere Torres et al. (1992) opgemerkt is, zal meer werk voor selectie van een voor een bepaald stroomgebied geschikte verliesfunctie nodig zijn voordat deze benadering volledig geimplementeerd kan worden.

9

References

Abbott, M.B., Bathust, J.C., Cunge, J.A., O'Connell, P.E. & Rasmussen, J. (1986), An introduction to the European Hydrological System-Systѐme Hydrologique Europḗen, "SHE", 2: Structure of a physically-based, distributed modelling system, Journal of Hydrology, 87, 61-77.

Abbott, M. B., and Minns, A. W. (1998), Computational hydraulics, Ashgate, Aldershot, 2nd. Edn., 557 pp.

ASCE Task Committee (2000a), Artificial Neural Networks in Hydrology. I: Preliminary Concepts, Journal of Hydrologic Engineering, Vol. 5, No.2

ASCE Task Committee (2000b), Artificial Neural Networks in Hydrology. II: Hydrologic Applications, Journal of Hydrologic Engineering, Vol. 5, No.2

Aizerman, M.A., Braverman, E.M. & Rozonoer, L.I. (1964 a), Theoretical foundations of the potential function method in pattern recognition learning, Automation and Remote Control, vol. 25, 204-215.

Aizerman, M.A., Braverman, E.M. & Rozonoer, L.I. (1964 b), The probability problem of pattern recognition learning and the method of potential functions, Automation and Remote Control, vol. 25, 1175-1193.

ANNEXG group, (2002), An international comparative study of artificial neural network techniques for rive stage forecasting, in Proceedings of the 8th National Hydrology Symposium, The Artificial Neural Network Experiment Group, British Hydrological Society, Birmingham, London, 75-81.

Bell, V.A., & Moore, R.J. (2000), Short period forecasting of catchment-scale precipitation. Part II: a water-balance storm model for short-term rainfall and flood forecasting, Hydrology and Earth System Sciences, 4(4), 627-633.

Beran, M. (1999), Hydrograph Prediction-How much skill?, Hydrological and Earth System Sciences, 3(2), 305-307.

Boogaard, H.F.P. Van den, Gautam, D.K., & Mynett, A.E. (1998), Auto-regressive neural networks for the modelling of time series, Hydroinformatics, Balkema, Rotterdam. The Netherlands, pp. 741-8.

Bowden, G.J., Maier, H.R. & Dandy, G.C. (2002), Optimal division of data for neural network models in water resources applications, Water Resources Research, 38(2), 1-11.

Box, G.E.P. & Jenkins, G.M. (1976), Time series analysis forecasting and control, Revised edition, Holden Day.

Bras, R.L., & Rodriguez-Iturbe, I. (1985), Random functions and hydrology, Addison-Wesley Publ. Co., Reading, Mass, 559 pp.

Bressand, F. (2002), The ALHTAIR project of the flood forecasting service from Gard, Houille Blanche, 2, 64-68.

Bruce, J.P. & Clark, R.H. (1966), Introduction to hydrometeorology, Pergamon Press, Oxford, 319 pp.

Cameron, D., Kneale, P. & See, L. (2002), An evaluation of a traditional and a neural net modelling approach to flood forecasting for an upland catchment, Hydrological Process, 16, 1033-1046.

Campolo, M., Andreussi, P. & Soldati,A., (1999, a), River flood forecasting with a neural network, Water Resources Research, 35(4), 1191-1197.

Campolo, M., Soldati, A. & Andreussi, P., (1999, b), Forecasting river flow rate during low-flow periods using neural networks, Water Resources Research, 35(11), 3547-3552.

Carriere, P., Mohaghegh, S., and Gaskari, R. (1996), Performance of a virtual runoff hydrograph system. Journal of Water Resource Planing And Management, ASCE, 122(6), 421-427.

Chapman, T, (1994), Comment on "How much complexity is warranted in a rainfall-runoff model?" by A J Jakeman and G M Hornberger, Water Resources Research, 30 (12), 3565.

Chow, V. T. (1959), Open-channel hydraulics, McGraw-Hill, New York, 680 pp.

Christensen, J.H. & Christensen, O.B. (2003), Climate modelling-Severe summertime flooding in Europe, Nature, 421 (6925), 805-806.

Cichocki, A., & Unbehauen, R. (1992), Neural Networks for Optimization and Signal Processing, John Wiley & Sons.

Dawson, C.W. & Wilby, R. (1998), An artificial neural network approach to rainfall-runoff modelling, Hydrology Sciences, Journal, 43, 47-66.

Dawson, C.W. & Wilby, R.L. (2001), Hydrological modelling using artificial neural networks, Progress in Physical Geography, 25(1), 80-108.

Deshons, P. (2002), Urban flood forecast and monitoring. Experience of Marseille city, Houille Blanche, 2, 56-59.

Dibike, Y.B. (1997), Use of Artificial Neural Networks in Model Approximation, M.Sc. Thesis HH 309, IHE-Delft.

Dibike, Y.B, Solomatine, D. & Abbott, M.B.(1999), On the encapsulation of numerical-hydroulic models in artificial neural network, Journal of Hydraulic Research, 37(2), 147-161.

Diskin, M.H., & Simon, E. (1977), A procedure for the selection of objective functions for hydrologic simulation model, Journal of Hydrology, 34, 129-149.

Dooge, J.C.I. (1959), A general theory of the unit hydrograph, Journal of Geophysic Research, 64(1), 242-256.

Dooge, J.C.I. (1973), The linear theory of hydrologic systems, US Dept. of Agriculture Tech. Bulletin, 1468.

Duan, Q., Sorooshian, S. & Gupta, V.K. (1994), Optimal use of the SCE-UA global optimization method for calibrating watershed models, Journal of Hydrology, 158, 265-284.

Duband, D., Obled, Ch. & Rodriguez, J.Y., (1993), Unit hydrograph revisited: an alternate iterative approach to UH and effective precipitation identification, Journal of Hydrology, 150, 115-149.

Durucan, S. & Imrie, C.E., (1998), The use of artificial neural networks in short-term river flow prediction, Proceedings of the Fifth International Symposium on Environmental Issues and Waste Management in Energy and Mineral Production. Ankara, Turkey, 243-248.

Elman, J.L. (1990), Finding Structure in Time, Cognitive Science, 14, 179-211.

Fahlman, S.E., Lebiere, C. (1991), The cascade-correlation learning architecture, Carnegie Mellon University Technical Report CMU-CS-88-162.

Fausett, L.V. (1994), Fundamentals of Neural Networks, Architectures, Algorithms, and Application, Prentice Hall, 372-385.

Flood, I., Kartam, N.(1994), Neural networks in civil engineering, I: principles and understanding, Journal of Computing in Civil Engineering, 8(2), 131-148.

Ford, D.T. (2001), Flood-Warning Decision-Support System for Sacramento, California, Journal of Water Resources Planning and Management, 127 (4): 254-260.

Franchini, M. & Lamberti, P. (1994), A flood routing Muskingum type simulation and forecasting model based on level data alone, Water Resource Research, 30 (7): 2183-2196.

Garrote, L. & Bras, R.L. (1995), A distributed model for real-time flood forecasting using digital elevation models, Journal of Hydrology, 167: 279-306.

Gustard, A., Roald, L.A., Demuth, S., Lumadjeng, H.S. & Gross, R. (1989), Flow Regimes from Experimental and Network Data (FREND), Volume II Hydrological Data.

Han, D. & Yang, Z. (2001) River flow modelling using support vector machines, Proc of Theme C, XXIX IAHR Congress, Beijing, China, September 16-21, 2001, 494-499.

Hall, M.J. (1977) The effect of urbanisation on storm runoff from two catchment areas in North London, in: Effects of Urbanisation and Industrialisation on the Hydrological regime and Water Quality, Proc. Amsterdam Sympo., International Association of Scientific Hydrology, No. 123, pp. 144-152.

Hall, M.J. (2001), How well does your model fit the data?, Journal of Hydroinformatics, 03,1, 49-55.

Hall, M.J. and Minns, A.W. (1993) Rainfall-runoff as a problem in artificial intelligence: experience with a neural network, Proc 4 th Nat. Hydrol. Symp., Cardiff, British Hydrological Society, London, pp. 5.51-5.57.

Hettiarachchi, P., Hall, M.J. & Minns, A.W. (2003), The extrapolation of artificial neural networks for the modelling of rainfall-runoff relationships, under review.

Hornik, K., Stinchcombe, M. & White, H. (1989), Multilayer feedforward networks are universal approximators, Neural Networks, 2: 359-366.

Hu, W. (2002), Flood Forecasting in the Wuding River Using Artificial Neural Networks and the Large Scale Precipitation Forecast, M.Sc. Thesis, IHE, Delft, The Netherlands, HH 441.

Hydrology & Earth System Sciences, (2000), HYREX: the Hydrological Radar Experiment, 4(4).

Imrie, C.E., Durucan, S. & Korre, A. (2000), River flow prediction using artificial neural networks: generalisation beyond the calibration range, Journal of Hydrology, 233, 138-153.

Jasper, K., Gurtz, J. & Lang, H. (2002), Advanced flood forecasting in Alpine watersheds by coupling meteorological observations and forecasts with a distributed hydrological model, Journal of Hydrology, 267 (1-2), 40-50.

Johnstone, D, and Cross, W P, (1949), Elements of applied hydrology, The Roland press, New York, 275 pp.

Jordan, M.I. (1986), Attractor Dynamics and Parrallelism in a Connectionist Sequential Machine. In Proceedings of the Eighth Annual Conference of the Cognitive Science Society (Amherst 1986), 531-546. Hillsdale: Erlbaum.

Jordan, M.I. (1989), Serial Order: A Parallel, Distributed Processing Approach. In Advances in Connectionist Theory: Speech, eds. J.L. Elman and D.E. Rumelhart. Hillsdale: Erlbaum.

Kaastra, I. & Boyd, M.S. (1995), Forecasting futures trading volume using neural networks, The Journal of Futures Markets, 15(8), 953-970.

Karunanithi, N., Grenney, W.J., Whitley, D., & Bovee, K. (1994), Neural networks for river flow prediction, Journal of Computing in Civil Engineering, 8(2), 201-220.

Kim, G & Barros, A.P. (2001), Quantitative flood forecasting using multisensor data and neural networks, Journal of Hydrology, 246: 45-62.

Kothyari, U.C. & Singh, V.P. (1999), A multiple-input single-output model for flow forecasting, Journal of Hydrology, 220: 12-26.

Lange, N.T.G. (1998), Advantages of unit hydrograph derivation by neural networks, Hydroinformatics'98, Babovic, V. & Larsen, L.C. (eds), Balkema, Rotterdam, 783-789.

Lardet, P. & Obled, C. (1994), Real-time flood forecasting using a stochastic rainfall generator, Journal of Hydrology, 162: 391-408.

Lee, Y.S. (1967), River forecasting techniques in Korea. In Hydrological Forecasting, Technical Note No.92, World Meteorological Organization No. 228 TP. 122,103-109.

Legates, D.R. & McCabe, G.J. (1999), Evaluating the use of "goodness-of-fit" measures in hydrologic and hydroclimatic model validation, Water Resource Research, 35 (1): 233-241.

Linsley, R.K., Kohler, M.A., & Paulhus, J.L.H. (1949), Applied hydrology, Mcgraw-Hill, New York, 689 pp.

Linsley, R.K., Kohler, M.A., & Paulhus, J.L.H. (1982), Hydrology for engineers, Mcgraw-Hill, New York, 3rd. edn., 508 pp.

Maier, H.R. & Dandy, G.C. (2000), Neural networks for the prediction and forecasting of water resources variables: a review of modelling issues and applications, Environmental Modelling & Software, (15), 101-124.

Mason, J.C., Price, R.K., and Tem'me, A. (1996), A neural network model of rainfall-runoff using radial basis functions, Journal of Hydraulic Research, vol. 34, 537-548.

Mellor,D., Sheffield,J., O'Connell, P.E. & Metcalfe,A.V. (2000), A stochastic space-time rainfall forecasting system for real time flow forecasting II: Application of SHETRAN and ARNO rainfall runoff models to the Brue catchment, Hydrology and Earth System Sciences, 4(4), 617-626.

Minns, A.W. (1998), Artificial Neural Networks as Subsymbolic Process Descriptors, Balkema, Rotterdam. The Netherlands.

Minns, A.W. & Hall, M.J. (1996), Artificial neural networks as rainfall-runoff models, Hydrological Science, Journal, 41, 399-417

Minns, A.W. & Hall, M.J. (1997), Living with the ultimate black box: more on artificial neural networks, Proc 6 th Nat. Hydrol. Symp., Salford, British Hydrological Society, London, pp. 9.45-9.49.

Minns, A.W. & Hall, M.J. (2003), Rainfall-runoff modelling, ch. 6 of Abrahart, R.J., See,L. and Kneale, P.E. (eds) Neural networks for hydrological modelling, Balkema, Rotterdam, in press.

Muttiah, R.S., Srinivasan, R., Allen, P.M. (1997), Prediction of two-year peak stream discharges using neural networks, Journal American Water Resource Association, (33), 625-630.

Nagai, A. (2002), Hydrologic Modelling of Rainfall-runoff process and Its Application to Real-time Flood Forecasting.(http://www.nourin.tsukuba.ac.jp)

Nash, J.E. (1957), The form of instantaneous unit hydrograph, Hydrological Science, Bulletin, 3, 114-121.

Nash, J.E. & Sutcliffe, J.V. (1970), River flow forecasting through conceptual models. Journal of Hydrology, 10, 282-290.

Nemec, J, (1986), Hydrological Forecasting. Design and Operation of Hydrological Forecasting Systems, D Reidel Publishing Company, Dordrecht, 239 pp.

NERC.(1975) The Flood Studies Report, Volume I-V, The Natural Environment Research Council, London.

Norgaard, M., Ravn, O., Poulsen, N.K., & Hansen, L.K. (2000), Neural Networks for Modelling and Control of Dynamic Systems, Springer, UK.

O'Donnell, T. (1985), A direct three-parameter Muskingum procedure incorporating lateral inflow, Hydrological Sciences Journal, 30 (4), 479-496.

O'Donnell, T., Pearson, C. P., & Woods, R. A. (1988), Improved fitting for three-parameter Muskingum procedure, Journal of Hydraulic Engineering, ASCE, 114 (5), 516-528.

Pedder, M.A., Haile, M.& Thorpe, A.J. (2000), Short period forecasting of catchment-scale precipitation. Part I: the role of Numerical Weather Prediction, Hydrology and Earth System Sciences, 4(4), 627-633.

Price, R. K. (1973), Flood routing methods for British rivers, Proceedings of the Institution of Civil Engineers, 55, 913-930.

Proaño, C.O., Minns, A.W., Verway, A., and Van den Boogaard, H.F.P., (1998), Emulation of a sewerage system computational model for the statistical processing of large numbers of simulations, Hydroinformatics, Balkema, Rotterdam. The Netherlands, pp. 1145-52.

Rumelhart, D.E., McClelland, J.L. & the PDP Research Group, (1986), Parallel Distributed Processing-Explorations in the Microstructure of Cognition, 2 Volumes, MIT Press, Cambridge, MA.

Schalkoff, R.J. (1997), Artificial Neural Networks. McGraw-Hill Companies, Inc

Sempere Torres, D., Rodriguez, J.Y. and Obled, CH. (1992), Using the DPFT Approach to Improve Flash Flood Forecasting Models, Natural Hazards 5: 17-41.

Sharma, K.D. & Murthy, J.S.R. (1995), Hydrologic Routing of Flow in Arid Ephemeral Channels, Journal of Hydraulic Engineering, 121 (6): 466-471.

Shamseldin, A.Y. (1997), Application of a neural network technique to rainfall-runoff modelling. Journal of Hydrology, 199, 272-294.

Sherman, L.K. (1932), Stream flow from rainfall by the unit-hydrograph method, Engineering News Record, 108, 801-805.

Shiiba, M., Laurenson, X. & Tachikawa, Y.,(2000), Real-time stage and discharge estimation by a stochastic-dynamic flood routing model, Hydrological Processes, 14, 481-495.

Sittner, W.T., Schauss, C.E. & Monro, J.C., (1969), Continuous hydrograph synthesis with an API-type hydrologic model, Water Resource Research, 5 (5), 1007-1022.

Singh, P.V. & Frevert, D.K. (2002a), Mathematical Models of Small Watershed Hydrology and Applications, Water Resources Publications, LLC.

Singh, P.V. & Frevert, D.K. (2002b), Mathematical Models of Large Watershed Hydrology, Water Resources Publications, LLC.

Snyder, F F, (1938), Synthetic Unit-graphs, Transections of the American Geophysical Union, 19, 447-454.

Somsin, N. & Varoonchotikul, P., (2002), Hourly flood forecasting of Chumporn river with a neural network model, Second International Symposium on Flood Defence, Beijing, Sep. 10-13, 2002.

Sugawara, M. (1995), Tank Model, in Ch 6 of Singh, V.P. (ed) Computer Models of Watershed Hydrology, WRP, Colorado, 165-214.

Takeuchi, K. (2001), Increasing vulnerability to extreme floods and societal needs of hydrological forecasting, Hydrological Sciences, 46 (6): 869-881.

Todini, E. (1996), The ARNO rainfall-runoff model, Journal of Hydrology, 175: 339-382.

Yang, Z. (2001), Flood routing in the River Huaihe based on the three-parameter Muskingum model, IHE Delft, MSc thesis HH 412, 91 pp.

Yu, X., Liong, S.Y. & Babovic, V. (2002), Hydrologic forecasting with support vector machine combine with chaos-inspired approach, Hydroinformatics, Cardiff, July 1-5, 2002, 764-769.

Varoonchotikul, P., Hall, M.J. & Minns, A.W. (2002a), Extrapolation management for Artificial Neural Network models of Rainfall-Runoff relationships, Hydroinformatics, Cardiff, July 1-5, 2002, 673-678.

Varoonchotikul, P., Hall, M.J. & Minns, A.W. (2002b), Flood Forecasting using Jordan Recurrent Artificial Neural Networks, Second International Symposium on Flood Defence, Beijing, Sep. 10-13, 2002.

Vink, G van der, et al, (1998), Why the United States is becoming more vulnerable to natural disasters, Eos, Transactions of the American Geophysical Union, 79 (44), 533-537.

Wheater, H.S., Boxall, S. & Wagener, T., (2002), Regionalisation of rainfall-runoff models: an application to the Thames Basin, BHS 8th National Hydrological Symposium, Birmingham, 199-206.

Wood, E.F. & O'Connell, P.E. (1985), Real-time forecasting, Ch. 15 of Anderson, M.G., and Burt, T.P., Hydrological forecasting, Wiley, New York, 505-558.

Wood, S.J., Jones, D.A. & Moore, R.J. (2000), Hydrology & Earth System Sciences, HYREX: the Hydrological Radar Experiment, 4(4), 531-543.

World Bank, (1993), Water Resources Management. A World Bank Policy Paper, The World Bank, Washington, DC, 140 pp.

Zealand, C.M., Burn, D.H. & Simonovic, S.P. (1999), Short term streamflow forecasting using artificial neural networks, Journal of Hydrology, 214, 32-48

Zhu, M.L., Fujita, M. and Hashimoto, N. (1994), Application of neural networks to runoff prediction, in Stochastic and Statistical Methods in Hydrology and Environmental Engineering, edited by K.W. Hipel et al., Kluwer Acad., Norwell, Mass., 205-216.

I

Data used for the study

Data from 7 river catchments were studied. All of them are listed in alphabetical order. The catchments varied from 23.99 to 1,819 km^2. One of them is located in Thailand and the rest are located in the United Kingdom. Each basin is described in the following subsections.

I.1. The Brue river

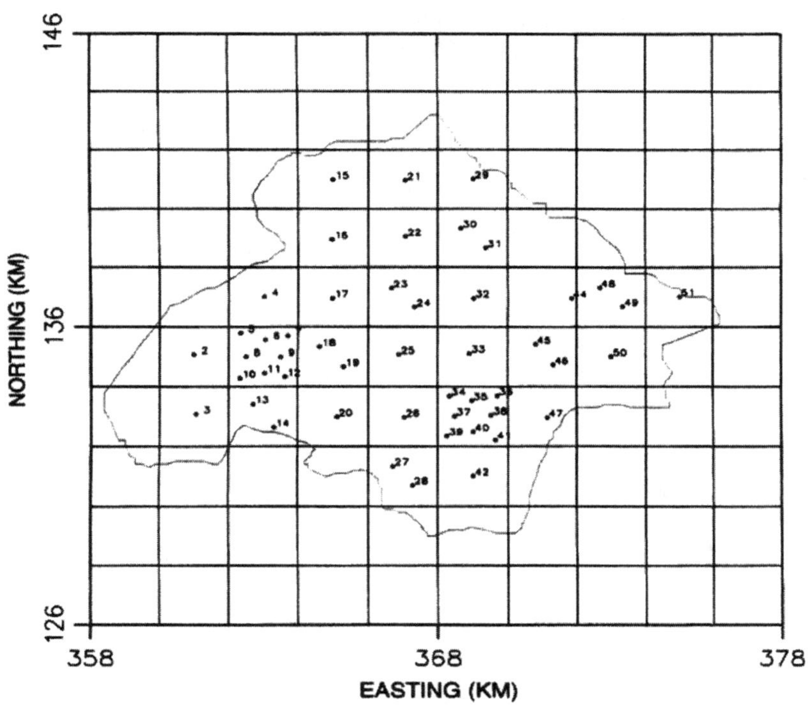

Figure I-1 The HYREX dense raingauge network over the Brue catchment. (Adapted from Wood et al,2000).

The Brue catchment of 135 km^2 is located at Lovington in the south west of the United Kingdom. From 1993 to 1997, there was a dense network of 49 automatic raingauges within the catchment in connection with the HYREX experiment (Hydrology & Earth System Sciences, 2000). This dense network of 49 tipping-bucket raingauges installed across the Brue catchment, in Somerset, South West England, is shown in Figure I-1. The catchment rises from an altitude of 20 m at the Crump weir gauging station to 260 m AOD on its eastern boundary. Land use is dominated by pastureland on clay soils with some patches of woodland in the higher relief eastern half of the catchment. Average annual rainfall for the standard period 1961 to 1990 was 867 mm. River flow has a mean of 1.88 m^3/s and since gauging began has reached an instantaneous record maximum of 96 m^3/s. The rainfall-runoff data used in this study were for the period 1st January 1994-31st July 1995. The automatic rainfall records were extracted at 15-min intervals, corresponding to the runoff data records.

I.2. The Chumporn river

The Chumporn river, located in the southern part of Thailand, is subjected to monsoon climate from August-December. The catchment area is shown in Figure I-2. The river basin at the gauging station named X.158 is 1,819 km^2, and includes two upstream gauging stations named X.46 and X.64. The station X.46 is located at Khlong Rap Ro and X.64 is at Khlong Tha Sae, Chumporn. Their areas are 751 and 957 km^2, respectively. The hourly hydrometers for 1990 were used for this study.

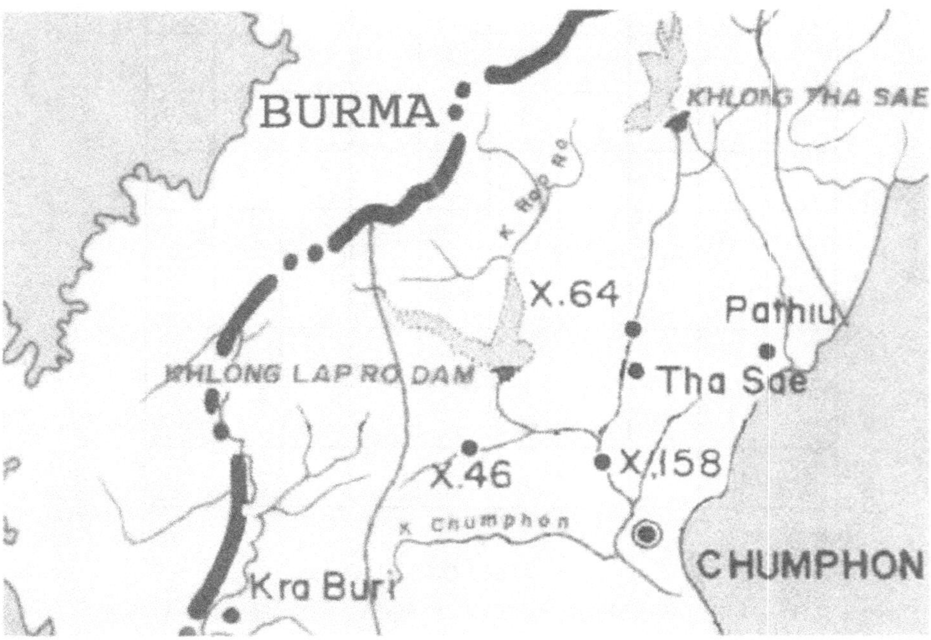

Figure I-2 The map of the catchment of Chumporn river, Thailand.

All the rainfall-gauging stations within the basin are equipped with totalisers and record as daily rainfall. Although the flood is generated by rainfall within the river basin, the rainfall data were not used in this study since the response time of the river basin is less than a day and rainfall data alone are not sufficient to compute flow (Minns & Hall, 1996). Therefore, daily rainfall will not match this characteristic of the basin. To overcome this deficiency, the two upstream hourly water level gauging stations were employed instead. The advantage of these runoff data is that the water levels upstream are already representative of the integrated effects of the rainfall in the upstream sub-area.

I.3. The Dart river

The Dart river is a catchment of 248 km^2, located in the southwestern part of the United Kingdom. Streamflow and rainfall records measured at hourly intervals were available for 22 events over the years 1963-69. The rainfall data were in the form of catchment average rainfall. These data were obtained from the Representative Basin Catalogue for Great Britain which maintained by the Centre for Ecology & Hydrology, UK.

I.4. The Dollis Brook river

The Dollis Brook is a catchment of 23.99 km^2 draining into the Brent Reservoir in north London. The geology of the area is predominantly London Clay, with outcrops of the Claygate beds and the Pebble beds occurring on the higher ground to the north. Since the 1920s, the catchment has also been extensively urbanised, such that the proportions of impervious area had by 1970 reached 21 per cent. Records of stream flows were available for Dollis Brook at Hendon Lane for 40 events over the period 1952-69. There are two autographic raingauges at Barnet and Mill Hill from which records were available during the period of discharge measurements. The map of the catchment is shown in Figure I-3. Further details are given by Hall (1977), who extracted the original data from recorder charts.

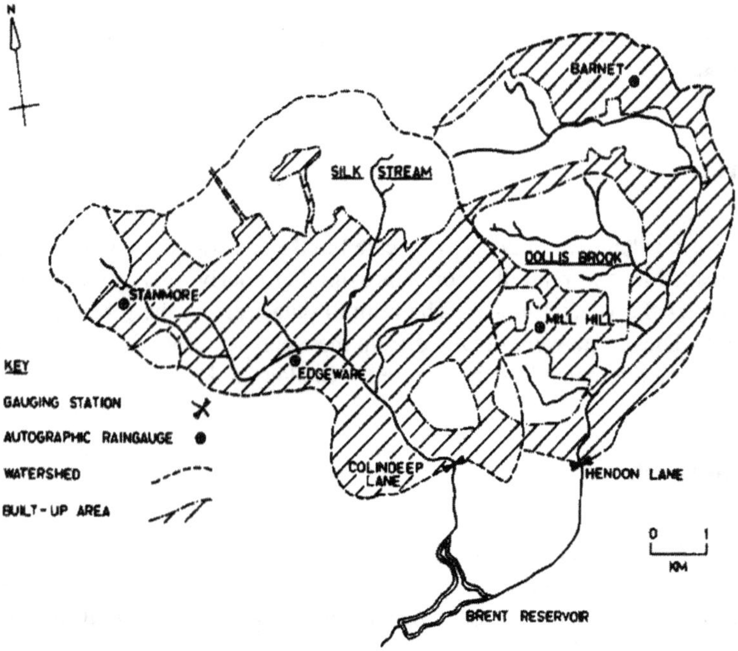

Figure I-3: The map of the catchment of Rivers Dollis Brook and Silk Stream (Adapted from Hall, 1977).

I.5. The River Mole

The River Mole at Kinnersley Manor is a lowland tributary of 142 km^2 of the River Thames, which drains a largely impervious catchment consisting mostly of Weald Clay. Flows are gauged using a rectangular flume, calibrated by current meter gauging which extend beyond bankfull. The England and Wales Environment Agency (Thames region) supplied rainfall-runoff data for the River Mole for the whole of 1994 (Dawson & Wilby; 1998). These data represented rainfall (mm) measured at 15-min intervals at the Burstow raingauge and 15-min flow measurements made at Kinnersley Manor (m^3/s). The river has experienced significant and increasing net imports of sewage effluent through time. The land use is very mixed with rural tracts and urban centres such as Crawley and Gatwick Airport. Average annual rainfall is 793 mm and river flow has an annual mean of 1.95 m^3/s and has reached an instantaneous record maximum 68.5 m^3/s.

I.6. The Silk Stream river

The Silk Stream is a catchment of 31.25 km^2 also drains into the Brent Reservoir in north London. The geology of the area is the same as the Dollis Brook. Since the 1920s, the catchment has been extensively urbanised, such that the proportions of impervious area had by 1970 reached 25 per cent. Records of stream flows were available for the Silk Stream at Colindeep Lane for 51 events over the years 1929-44. There are two autographic raingauges at Stanmore and Edgeware from which records were available during the period of discharge measurements. The map of the catchment is also shown in Figure I-3. Further details are given by Hall (1977), who extracted the original data from recorder charts.

I.7. The Thrushel river

The Thrushel river is a catchments of 112.7 km^2, also located in the south-western part of the United Kingdom. Streamflow records at hourly interval were available for 37 events over the years 1971-85. The rainfall data were in the form of catchment average rainfall. These data were obtained from the Representative Basin Catalogue for Great Britain which maintained by the Centre for Ecology & Hydrology, UK.

I.8. Esimated Maximum Flood (EMF) of the Dart and Thrushel rivers

The *Estimated Maximum Flood* (EMF) might be defined as the flood with a very small, but non-negligible, probability of annual exceedance. It is estimated by a procedure that is an extension of the unit hydrograph plus design storm approach in the UK Flood Studies Report (NERC, 1975). In this method, the unit hydrograph and the loss function were derived from a regional analysis of UK catchments, and the PMP storm profile is based upon the extrapolation of UK rainfall statistics. This type of regional domain knowledge was applied to Dart and Thrushel rivers, the relevant EMFs being 1,867 and 760 m3/s, respectively.

Curriculum vitae

Personal Information

Full Name	:	Pichaid Varoonchotikul
Year of Birth	:	1957
Place of Birth	:	Bangkok, Thailand
Profession	:	Civil engineer
Nationality	:	Thai

Academic Background

Bachelor of Engineering in Civil Engineering in 1980 from Chulalongkorn University

Master of Science in Environmental Engineering in 1984 from the University of Michigan, USA.

Master of Science with distinction in Hydrological Engineering in 1989 from the International Institute for Infrastructural, Hydraulic and Environmental Engineering (IHE) in Delft, The Nederlands

Professional Experience

From 1985-1996 : Hydrologist, The Royal Irrigation Department, Thailand.

Responsibilities: Planned and designed Irrigation structures.

From 1997-1999 : Lecturer, Srinakarinwiroj University, Thailand.

Responsibilities: Teaching basic and professional courses in civil engineering related subjects, such as, Timber and steel desigh, strength of material, Numerical method.

Publications

1. Varoonchotikul, P., Hall, M.J. & Minns, A.W., (2002), Extrapolation management for Artificial Neural Network of Rainfall-Runoff relationships, Hydroinformatics, Cardiff, July 1-5, 2002, 673-678.
2. Varoonchotikul, P., Hall, M.J. & Minns, A.W., (2002), Flood Forecasting using Jordan Recurrent Aritificial Neural Network, Second International Symposium on Flood Defence, Beijing, Sep. 10-13, 2002.

For Product Safety Concerns and Information please contact our EU
representative GPSR@taylorandfrancis.com
Taylor & Francis Verlag GmbH, Kaufingerstraße 24, 80331 München, Germany

www.ingramcontent.com/pod-product-compliance
Lightning Source LLC
LaVergne TN
LVHW021113300926
849160LV00009B/720
9789058096319